BASIC Electronics

David Parsons

MACMILLAN

Acknowledgements

The author and publishers wish to thank the following who have kindly given permission for the use of copyright material:

UNILAB of Blackburn for their considerable help providing photographs of their excellent apparatus;
BICC VERO of Southampton for authorising pictures of their circuit boards and wiring devices;
Lowe Electronics Ltd of Matlock, for the picture of the HF125 receiver;
MK Electric Ltd for pictures of electrical devices;
Maplin Electronics of Rayleigh for permission to mention their catalogue;
Cirkit Distribution Ltd of Broxbourne;
Rapid Electronics of Colchester for permission to quote their catalogue;
Everyday Electronics magazine;
Professor D.G. Harvey for photographs taken in Manhood Community College, Selsey;
Heinemann Professional Publishing for the diagram on page 82;
Liz Drake, for the sketch on page 83.

Thanks are also due from the author to:

Tony Walsh and Ian Parsons for very valuable help in script reading and for suggesting improvements;
his wife, Jan, for typing the manuscript and for her continued support, help and encouragement.

The publishers have made every effort to trace the copyright holders, but if they have inadvertently overlooked any, they will be pleased to make the necessary arrangements at the first opportunity.

© David Parsons 1988

All rights reserved. No reproduction, copy or transmission of this publication may be made without written permission.

No paragraph of this publication may be reproduced, copied or transmitted save with written permission or in accordance with the provisions of the Copyright Act 1956 (as amended), or under the terms of any licence permitting limited copying issued by the Copyright Licensing Agency, 33–4 Alfred Place, London WC1E 7DP.

Any person who does any unauthorised act in relation to this publication may be liable to criminal prosecution and civil claims for damages.

Published by
MACMILLAN EDUCATION LTD
Houndmills, Basingstoke, Hampshire RG21 2XS
and London
Companies and representatives throughout the world

Produced by AMR for
Macmillan Education Ltd

Printed in Hong Kong

British Library Cataloguing in Publication Data
Parsons, David
  Basic electronics.
  1. Electronics
  I. Title   II. Series
  537.5
  ISBN 0–333–46789–2

Contents

Introduction	3
Circuits	5
Light-emitting Diodes	7
Practical Ways of Joining Components and Circuits	9
Series and Parallel Circuits	13
Using Ammeters	14
Measuring Voltage	17
Resistance	20
Resistors	24
What is Electricity?	28
The Cathode Ray Oscilloscope	31
Rectifying AC to Produce DC	34
Capacitors	37
Electrical Power	40
Transformers	43
More about Diodes	45
Transistors	49
Other Ways of Switching on Transistors	52
Transducers and Relays	54
Fault Finding	58
Amplifiers	61
Integrated Circuits	63
Logic Systems	65
Microcomputer Systems	70
Audio Systems	74
Radio Receiver Systems	77
Feedback Control Systems	80
Sources of Danger	84
Fuses and Residual Current Devices	88
Power Supplies, Cable Sizing, Device Handling	90
To Sum Up	93
Components Required	95
What next?	95
Answers	96

Introduction

This book forms part of a Macmillan series designed for use by a wide range of students in school and further education, to provide practice in the development of core skills and knowledge in a variety of essential areas.

In particular, the series aims to ensure competence in core skills so that potential employees – school and college leavers – actually master the basic skills required by employers. Many of the skills included in the series are also needed in everyday life.

The approach taken is to present information in clear and carefully controlled steps and to provide numerous straightforward questions and tasks designed to test skills and explore the information presented.

The books are suitable for use not only with those pupils and students normally expected to take GCSE examinations, but also with those at all levels of ability, including students on pre-vocational courses such as CPVE and RSA Vocational Preparation. Each book is based upon the syllabus of the basic skills tests of the Associated Examining Board (Stag Hill House, Guildford, Surrey GU2 5XJ).

The series does not provide a framework for GCSE courses, but the books can be used in connection with a recognised public examination. They can also be used to provide practice in the core competences in the curriculum for the 14–19 age group.

Basic Electronics is designed to develop a knowledge and understanding of this most interesting subject. The content provides a broad introduction to the main areas and covers the topics specified in the AEB syllabus.

It is hoped that it will be of use, not only to those taking the AEB examination, but also to that large number of people who are interested in the subject as a leisure activity.

No previous knowledge of electricity or electronics is assumed.

Practical activities are included in 22 sections. Many involve making up circuits – including amplifiers and a radio, conducting experiments and using measuring instruments. These circuits have been designed to give satisfactory results using the minimum of components. Over 140 questions are included for self testing. The main topic areas covered include

- components such as transistors and integrated circuits (silicon chips) and circuit construction
- electronic systems including digital circuits, audio and radio systems
- the use of instruments for measurement and fault finding
- electrical safety.

The BSI are introducing some new component symbols and these have been included in the book as alternatives.

Circuits

In the picture are five everyday things which work by electronics. Some are very simple, like the torch. Others are much more complicated. But they all use *devices* or *components* arranged into *circuits*. The torch uses a bulb, a battery, a switch and metal strips to connect these into a circuit. In this torch the battery has two cells.

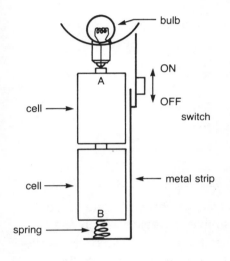

This diagram shows the important parts. The metal strip *conducts* electricity from the battery to the bulb. *Metals* conduct electricity. Copper, silver and gold are the best conductors. Most *non-metals* are *non-conductors* or *insulators*.

The end of the cell labelled A touches one end of the bulb. The end of the cell labelled B is connected to the bulb by the spring, metal strips and switch.

The diagram of the torch is not very easy to draw. It is much easier to draw a *circuit diagram* or *circuit*. In electronics circuit diagrams are always used. The circuit diagram just has the essential details. The circuit diagram for this torch would look like this.

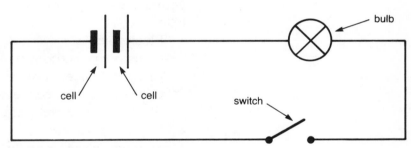

Circuit diagram for the torch

In a circuit diagram, we use *symbols* for each device or component. You will find it easy to learn each symbol as you come to it in this book. There is a list of all the components and symbols at the end of the book.

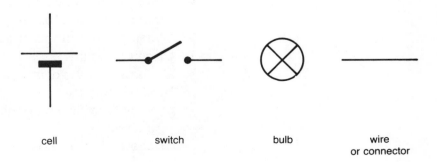

5

Task

Copy this summary into your notebook, filling in the blanks. Do *not* write in this book.

Circuits
Electronic things like calculators and torches use components arranged in c...s. The ... is carried along wires. The wires are made of ... because metals ... electricity. Non-metals do not conduct; they are .... In electronics we use circuit d..... Components are shown by s....

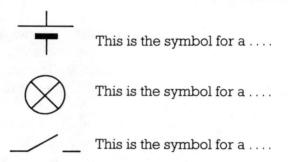

This is the symbol for a ....

This is the symbol for a ....

This is the symbol for a ....

Questions

The questions below are *multiple choice*. Only one of the answers given is correct. You have to decide which one. Check your answer with the list at the end of the book. Write your answer in your notebook. You need only write the letter.

1. Electronics
   a) uses circuits connected into symbols
   b) uses devices connected into circuits
   c) uses symbols connected into circuits
   d) always uses complicated systems.
2. Conductors are
   a) metal wires – usually copper – which carry electricity
   b) non-metals which insulate
   c) parts of circuits which make electricity
   d) wires made of things like glass or rubber.
3. Which of these is the symbol for a cell?

   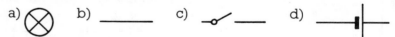

4. Which of the symbols in question (3) represents a switch?
5. Which of the symbols in question (3) represents a bulb?
6. Which of the symbols in question (3) represents a wire?
7. Circuit diagrams
   a) show only the components or devices in a circuit
   b) show only the conductors in a circuit
   c) show how the components are connected by wires in a circuit
   d) show a large amount of detail which is not essential.

Light-emitting Diodes

Modern electronic circuits use *light-emitting diodes* or *LEDs*. They glow with a red or green light and are often used as indicators.

This is what an LED looks like   **This is its symbol**   **or**

This is what a resistor looks like

This is its symbol

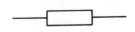

To make an LED light up you need a *4.5 volt battery*. '4.5 volts' tells us something about the battery. We will learn more about this later.

You also need a *resistor* with a value of *330 ohms*. This will have four coloured bands on it. They are orange, orange, brown and either gold or silver. These colours indicate the value of the resistor.

The resistor stops too high an electric *current* from passing through the LED.

Task

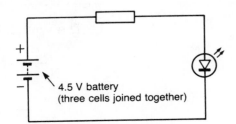

Try to connect up a circuit from this circuit diagram. Use lengths of wire with crocodile clips on the ends to connect it all.

If it *does not* work connect *either* the battery *or* the LED the other way round. If it *does* work, reverse *either* the battery *or* LED and it will go out.

Both the battery and LED have ends which are different. We say they have *polarity*. The battery has one terminal marked + which means *positive* and the other terminal is marked − which means *negative*. The long lead or *anode* on the LED has to be connected to + and the short lead or *cathode* has to be connected to −. (*Beware*, if you are using old LEDs. Someone may have cut bits off the leads. They may still work, but the leads will not be as shown. Try changing them round to find out.)

The LED conducts and lights up when the anode is connected to + and the cathode is connected to −. We say it is *forward biased*. The LED does *not* conduct or light up when the anode is connected to − and the cathode is connected to +. We say it is *reverse biased*. This experiment should show you something else about circuits. Try disconnecting it at any point. It stops immediately. This is because a circuit has to be *complete* in order to work. There must be a route for the electricity *all the way round* from the battery and *back* to the battery. If there is a gap anywhere there is not a complete circuit and it will stop working.

Try connecting the resistor the other way round. Does it make any difference? It should not. Resistors do not have polarity. They conduct either way.

Task

Copy this summary into your notebook, filling in the blanks.

Light-emitting diodes
Electrical circuits must be . . . in order to work.
If there is a gap they do not . . . . LEDs glow when the . . . is connected to + and the cathode is connected to −. The resisitor stops the . . . rising too high in the . . . . In an LED the . . . lead is the anode and the short lead is the . . . .

Questions

1. The symbol for an LED is

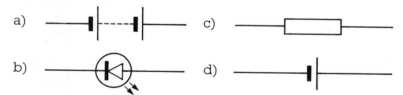

2. Which of the symbols in question (1) shows several cells joined together?
3. Which of the symbols in question (1) shows a resistor?
4. An LED
    a) conducts and glows when it is forward biased with the anode connected to −
    b) does not conduct when it is reverse biased and its anode is connected to +
    c) conducts and glows when it is forward biased with the anode connected to +
    d) will work whichever way it is connected.
5. The resistor which is connected to the LED
    a) controls the battery voltage
    b) is there only to add extra colours to the whole thing
    c) increases the current to a high value
    d) stops the current rising to a value where it might damage the LED.

Practical Ways of Joining Components and Circuits

Crocodile clips

We may use *wires with crocodile clips*, as in the arrangement in the last section. This is a quick and convenient way for very simple circuits. But for more complicated circuits it becomes difficult and is very liable to go wrong.

Breadboards

For more complicated circuits *plug-in breadboards* are better. These are plastic blocks which have holes in the top, into which component leads fit. In the holes are springy metal contacts which automatically connect components to all the other contact points in *that row only*. There are several commercial types (such as *Verobloc*).

Below is a photograph of a breadboard and a diagram of the way the contacts connect with each other. Study them carefully.

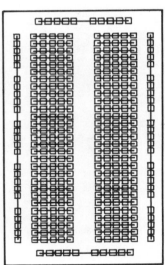

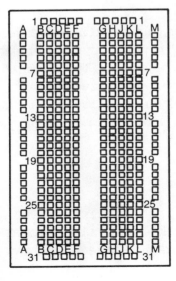

All the contacts in column A connect with each other, but with nothing else. All the contacts in column M connect with each other but with nothing else.

All the contacts in row 1 connect with each other. All the contacts in row 31 connect with each other.

Row 2, which does not have a number printed on it, has five contacts on the *left* which connect with each other, and five contacts on the *right* which connect with each other. There is *no* connection between contacts across the central division. F does not connect with G. There are 29 rows of these contacts in columns B, C, D, E, F and G, H, J, K, L.

To connect a component to others in that row all you need to do is to push the end of its wire lead into one of the holes. If the end of its wire lead is bent you will need to straighten it with pliers first.

Task

Assemble an LED and its resistor on a breadboard, using the circuit on page 7.

You will find it much easier if you follow some simple rules when using breadboards.

- Connect column A to positive on the battery.
- Connect column M to negative on the battery.
- Make connections from these + and − supplies to the rows of five, using small lengths of wire.

Position the LED with one lead in any of the short rows 2 to 30 and its other lead in any *other* row. Now do the same with the resistor. Connect the resistor to the LED. Finally put in the power supply connections and make the LED light up.

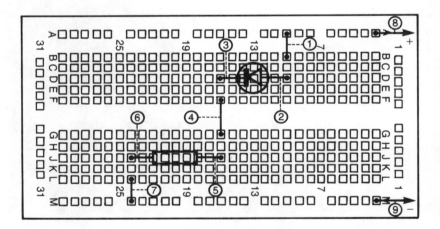

The diagram shows all the steps, numbered in sequence. When you have done this take it to pieces and assemble the same circuit using a different arrangement. The arrangement of the circuit is called a *layout*. Very many layouts are possible with just this one circuit on a breadboard. Try some out. The photograph shows the LED, resistor and battery connected on a Verobloc.

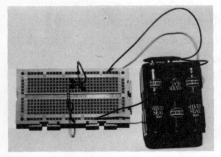

If you want to try drawing out layouts on a breadboard then copy the breadboard diagram on page 9 and draw on your copy.

Wiring boards

A more permanent way of connecting components is to *solder* them onto *wiring boards*. These have copper strips underneath which connect wires in rows, just like breadboards.

To solder a component to the board take these steps.
1. Plug in the soldering iron and let it warm up. (**Take care. It becomes very hot.**)
2. Push the component lead through the hole in the board and bend it over.
3. Hold the soldering iron on the end of the lead where it touches the copper track. Touch the solder onto the lead and track so that a small amount melts. Remove the iron, let the joint cool.

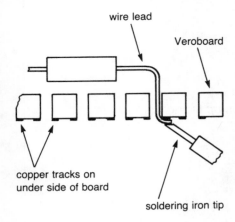

Precautions to take when soldering

- The end of the iron and parts to be soldered must be clean.
- Put the solder on the heated joint. Do *not* put it on the iron.
- The solder should 'wet' the conductors you wish to join. A soldered joint which has failed is called a *dry joint* because the solder has not wetted the conductors. Dry joints are unreliable connections.
- Take care not to touch components themselves with the iron. Do not overheat components by holding the iron on them for too long. They may be damaged by overheating.
- A *hot* soldering iron, *quickly applied* to *clean* connections which are in *close contact* is the best method of soldering.

Good reliable soldering takes time to learn. If you spend time practising it you will learn a useful skill.

Wiring pens

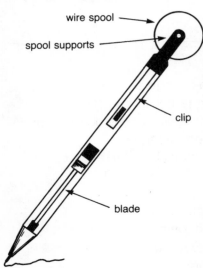

Another way of connecting components is to use a *wiring pen* system – like the easiwire tool made by BICC VERO.

You plug components into one side of a plain board which is perforated with holes. On the other side of the board you cut the component leads off so that about 3 mm remains. Using the Easiwire tool, you wind the wire round one of these projecting leads. Then you move the pen to the place where you want to connect that wire. The wire automatically feeds out as you go. You wind the wire round the next component lead, finally cutting it off. At each component lead it is best to wind the wire first *up* then *down*. This makes sure that the wire is close to the board.

The advantages of this system are that it is stronger than a breadboard yet it is still very easy to make changes in the circuit. You may re-use components and short leads do not matter.

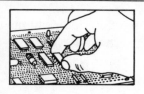
- Insert the components through the holes in the board.

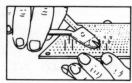
- Cut the leads off to a length of about 3 mm.

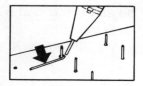
- Hold the wire from the pen.

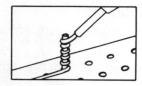
- Wind the wire four or five times *up* round the lead.

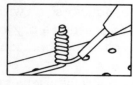
- Wind the wire *down* the lead.

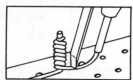
- Move the wire to the lead to which it is to be connected. Repeat the wrapping. Cut the wire. The two component leads are now joined.

Task

Copy this summary into your notebook, filling in the blanks.

Breadboards
Breadboards allow ... to be set up quickly. ... may be taken out and reused. ... is used for circuits which are needed permanently.

Questions

1 Wiring boards
 a) enable components to be changed easily
 b) enable circuits to be changed easily
 c) are as good as circuits with leads and crocodile clips
 d) have components soldered into them.
2 Which of the answers to question (1) is correct for a breadboard?
3 When soldering
 a) apply the solder to the iron
 b) apply solder to the heated joint
 c) make sure the solder does not wet the joint
 d) do not have the iron too hot.

Series and Parallel Circuits

There are two common ways of arranging components in circuits. These are *series* circuits and *parallel* circuits. Each produces different results. Some components may be arranged either way. A few things, such as meters, must be used in only one arrangement, *either* series *or* parallel.

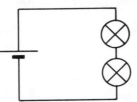

These bulbs are in series

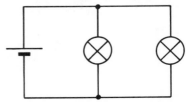

These bulbs are in parallel

Series circuits

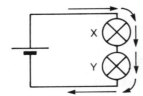

In a *series* circuit the current flows through one bulb before it goes to the next. There is only one route.

What do you think would happen if you unscrewed bulb X? Would Y remain alight? (**Hint** If X is unscrewed would there still be a complete circuit?)

Parallel Circuits

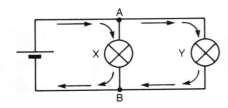

In a *parallel* circuit the current splits up at A. Some current goes through each bulb. Then the two currents rejoin at B.

What do you think would happen if bulb X was unscrewed in this circuit? Would Y remain alight?

Tasks

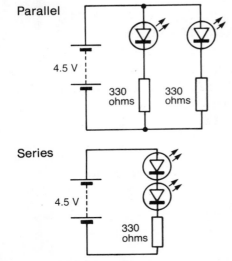

1. Try making up each of these LED circuits on your breadboard.

   Hints
   - Connect the long column A on the breadboard to the battery + and the long column M to the battery − Use short pieces of wire to connect the LEDs to the rows.
   - In a circuit symbol for a battery, the long line is always the positive + terminal.

   Now try disconnecting LED 1 in each circuit by just pulling it out. What happens to LED 2? Is this what you expected from thinking about bulbs X and Y? In a parallel circuit when LED 1 is removed LED 2 stays alight. In a series circuit, removal of LED 1 puts out LED 2.

2 Copy this summary into your notebook, filling in the blanks.

Series and parallel circuits
In a series circuit the current has only ... route through the components. In a ... circuit the ... may divide and go along different routes.

3 Copy and complete these circuit diagrams to show two bulbs a) in series, b) in parallel.

Questions

1 When bulbs are in series
   a) there is no need for a complete circuit
   b) if one is taken out the rest remain alight
   c) if one is taken out then they all go out
   d) the circuit needs more insulation.
2 Which of the answers to question (1) is correct for bulbs in parallel?

Using Ammeters

In this section we will learn how to use an *ammeter* to measure *current*.

Measuring current

The electric *current* which flows in a circuit is measured in *amperes*, usually called *amps*. In electronic circuits the current may be very small, so we also use the *milliamp*, which is one thousandth of an amp. A car headlight bulb might usually pass a current of 5 amps, written 5 A. A torch bulb may pass 0.3 A or 300 milliamps, written 300 mA. An LED may pass 20 milliamps (20 mA). The actual currents will vary with different components.

> **Remember**
>
> *current* is measured in *amps* (A) and in *milliamps* (mA)
>
> 1 mA is $\frac{1}{1000}$ of an amp or 0.001 A.
>
> 1000 mA = 1 A

Using a Multimeter

Multimeters are used to measure different electrical quantities. The photograph shows some types. You select the quantity you need to measure by using the switch.

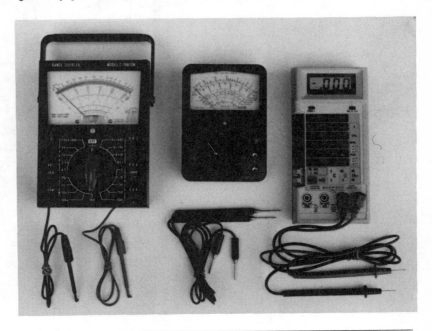

Task

This is the symbol for an ammeter

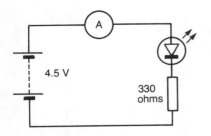

You are going to use a meter to measure the current in an LED circuit. The components will be in *series*. You are going to use a multimeter as an *ammeter*. Examine the one you are going to use very carefully. Plug the red lead into the socket marked + and the black lead into the − socket (on some meters + may be marked 'VΩA', and − may be marked 'com').

Set the switch to the highest value of DC mA or just 'A' if that is all your meter shows. **Do not connect it to anything yet**. An ammeter can be damaged very easily – follow these instructions carefully.

Wire up this circuit on your breadboard. Make sure the meter is connected as shown. (Get someone else to check it before connecting the battery if you have any doubts.) Check that the LED lights.

You should now be able to read off the value of the current flowing. You may need to spend a little time deciding which row of figures to read, on some meters. If the meter needle goes to the left, reverse the meter connections. If the needle moves only a very small way you may switch the meter to a lower range. This gives greater sensitivity.

The circuit arrangement above is a series circuit. Ammeters are **always** used in series, with *at least* one other component. **Never, ever,** on any account whatsoever, connect an ammeter directly to a battery, or in parallel with another component. You would wreck the meter.

Tasks

1 Put another LED in the circuit, like this.

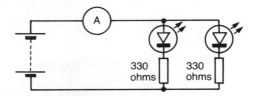

What is the current now? You should find that it rises to about twice its previous value. The LEDs are in parallel, but the *meter* is still in *series* with both LEDs. You may add a third LED if you wish.

2 Copy this summary into your notebook, filling in the blanks.

Ammeters

Ammeters measure the ... flowing in a circuit.
The unit of current is the .... The amp is quite large so the ... is also used and is one ... of an amp. Ammeters are a...s used in s... with other components.

Questions

1 A milliamp is
   a) 100 amps   b) 0.1 amps
   c) 0.01 amps   d) 0.001 amps.

2 Ammeters
   a) are always used in parallel with LEDS
   b) are always used in series with other components and may easily be damaged
   c) are never used in series
   d) are used in series and cannot easily be damaged.

3 Which of the following circuits shows an ammeter used incorrectly?

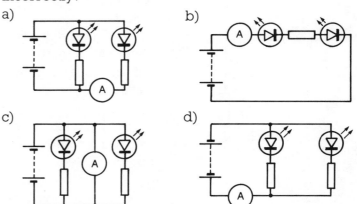

4 In which of the circuits in question (3) will the LEDs definitely not light?
5 In which of the circuits in question (3) will the meter show the current through only one LED?
6 Which of the circuits in question (3) shows LEDs in series?

Measuring Voltage

Cells in series: voltages across resistors in series

We can use the multimeter to measure another electrical quantity called *volts*, or *voltage*. The voltage is simply a measure of the electrical force which is driving the current round the circuit. In fact the term *electro-motive force* (or *EMF*) is more correct but the term voltage is very commonly used in practical situations.

Examples of voltages are

- a single dry cell, 1.5 volts
- a car battery, 12 volts
- mains electricity, 240 volts
- a nickel cadmium rechargeable cell, 1.2 volts.

It is important to understand the difference between *current* which is measured in *amps* and *EMF* which is measured in *volts*. The current, or number of amps flowing, is the quantity of electricity flowing along a wire each second. The EMF or voltage is the force driving that current along the wire.

To measure volts using your multimeter set the range switch to DC volts or just DCV. You will find a number of ranges available. Typical ones are 1000 V, 500 V, 50 V, 10 V.

The only safety rules to observe with a multimeter reading volts are:

- High voltages are dangerous. Do not touch them.
- If you are unsure of a voltage you are measuring, start with the meter set to a high range, say 500 V, then reduce the setting until a clear reading is obtained.

Time spent reading actual voltages will make the process clear. Voltmeters are *always* used in parallel. The only way they can be damaged electrically is by feeding them too high a voltage (for example, by setting it to 10 V and connecting it to 500 V – **don't try it**).

This is the symbol for a voltmeter

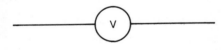

Task

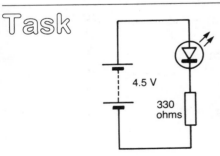

Set up this circuit on a breadboard.

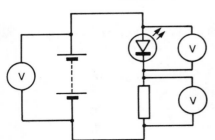

Measure the voltages across each component as shown.

17

You may measure the voltage of cells by simply connecting the meter across them. Try this.

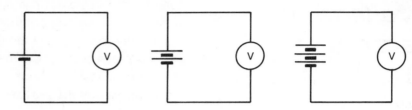

You should find that the total voltage of cells in series is the sum of the separate voltages; i.e. if one cell is 1.5 V two will be 3 V and so on.

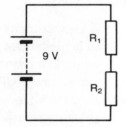

Now arrange this circuit on your breadboard. The resistors are named R_1 and R_2. Use 1000 ohm resistors for both R_1 and R_2.

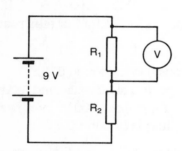

Measure the voltage across each resistor by connecting the meter like this. You should find that the voltage across each resistor is half the battery voltage – about 4.5 V.

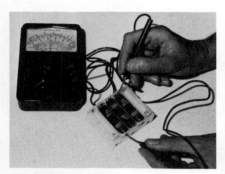

Measuring voltage

Now remove R_2 and replace it with a 2200 ohm resistor. Measure voltages across both resistors again. You will find that the voltage across R_1 is now *about* 3 volts and the voltage across R_2 is about 6 volts. Now make R_1 2200 ohms and R_2 1000 ohms. What do you expect the voltage to be? Check with your meter.

The voltages across resistors in series divide up according to the values of the resistors. The larger resistor will have the higher voltage across it. This circuit is called a *potential divider*. It is very useful for getting the correct voltages for making transistors work.

Some mathematics

To calculate the voltage across a resistor we can use this formula:

$$\text{voltage across } R_1 = \frac{\text{value of } R_1}{R_1 + R_2} \times \text{battery supply voltage}$$

In symbols $\displaystyle V_{R_1} = \frac{R_1}{R_1 + R_2} \times V_{\text{bat}}$

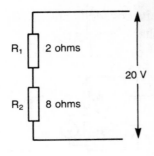

Examples
- Find the voltage across R_1.

$$V_{R_1} = \frac{R_1}{R_1 + R_2} \times V_{bat}$$

$$V_{R_1} = \frac{2}{2+8} \times 20$$

$$V_{R_1} = \frac{2}{10} \times 20$$

$$V_{R_1} = \frac{40}{10}$$

$$V_{R_1} = 4 \text{ volts}$$

You may use a calculator to help work out the answer.

- Find the voltage across R_2.
  The supply voltage is 20 volts.
  We have already worked out V_{R_1} to be 4 volts.
  So V_{R_2} must be $20 - 4 = 16$ volts.

You could work it out the long way if you want to.

$$V_{R_2} = \frac{R_2}{R_1 + R_2} \times V_{bat}$$

$$V_{R_2} = \frac{8}{2+8} \times 20$$

$$V_{R_2} = \frac{8}{10} \times 20$$

$$= 16 \text{ volts}$$

Task

Copy the summary below into your notebook. Fill in the blanks.

Voltmeters
Voltmeters are always connected in ... with a component. They measure the electrical force or EMF which drives the ... round the .... When measuring an unknown voltage always start with the meter set to the ... range. The voltage of three 1.2 cells in series would be ... volts. The symbol for a voltmeter is ....

Questions

1 Voltmeters
   a) measure amps and are used in parallel
   b) measure amps and are used in series
   c) measure volts and are used in parallel
   d) measure volts and are used in series.

2. Which of the following circuits correctly shows a voltmeter measuring the voltage across R_2?

a)   b)   c)   d)

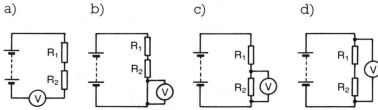

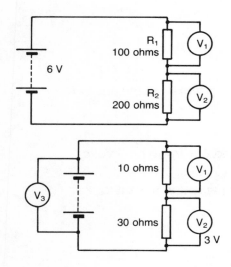

3. In this circuit
   a) V_1 will be 2 volts and V_2 will be 4 volts
   b) V_1 will be 4 volts and V_2 will be 2 volts
   c) V_1 and V_2 will each be 3 volts
   d) V_1 and V_2 will each be 6 volts.

4. In this circuit, the voltage V_2 across the 30 ohm resistor is 3 volts
   a) V_1 will be 10 volts and V_3 will be 4 volts
   b) V_1 will be 1 volt and V_3 will be 3 volts
   c) V_1 will be 1 volt and V_3 will be 30 volts
   d) V_1 will be 1 volt and V_3 will be 4 volts.

If you do not understand the calculations or potential dividers straight away, don't worry. Have a second and third try at the questions in a day or so and you will probably find you understand them easily.

Resistance

The relation between volts, amps and resistance

On page 7 we used a 330 ohm resistor. This resistor limited the current flowing in the LED. We have also met current, measured in amps, and electromotive force, measured in volts. Before going any further, make quite sure in your mind you understand the difference between current and electromotive force, between amps and volts. If you *do* understand, then you will find the next section easy. Re-read this paragraph and the last two sections if you are unsure.

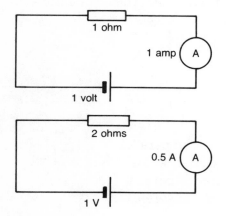

A resistor is a component which reduces the current flowing. Its value is measured in *ohms*. A resistor with a value of *1 ohm* will allow *1 amp* to flow through it when it is connected to *1 volt*.

Would the current be greater or less if you replaced the 1 ohm resistor with one of 2 ohms? Obviously it should be less – half an amp in fact.

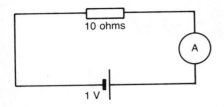

What would the current be if you now put in a 10 ohm resistor? The answer is one tenth of an amp (or 0.1 A).

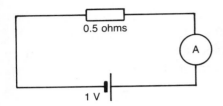

If you had a ½ ohm resistor what would the current be? You should have found that the current will be larger than 1 amp. It will be 2 amps.

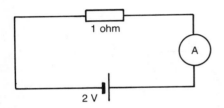

The smaller the resistor, the greater is the current. The larger the resistor, the less is the current.

So far we have only had a 1 volt cell in the circuit. Suppose we have a larger one with the original 1 ohm resistor. Since the electromotive force (or voltage) is now greater, more current is driven round the circuit. The voltage has been doubled so we would expect double the current to flow. That is in fact exactly what happens and we should find 2 amps flows.

Question

Examine these circuits carefully and decide what current flows.

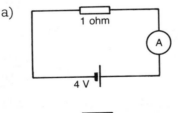
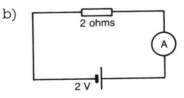
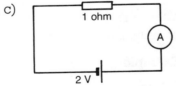
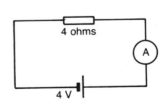

If you think about it carefully you should be able to work out that in a) the current is 4 amps. In b) there is twice as much resistance as in a), but there is twice the voltage too. Twice the resistance means half the current but twice the voltage means twice the current – so the answer is 1 amp again. 2 amps will flow in c) and 1 amp in d).

By this time you may have realised that the current flowing is the voltage *divided by* the resistance.

$$\text{current in amps} = \frac{\text{volts}}{\text{resistance in ohms}}$$

In symbols $I = \dfrac{V}{R}$

(I represents the current, in amps.)

Example

We can use this formula to calculate the current flowing in these circuits. Don't look at the solutions underneath until you have worked them out. A calculator may help.

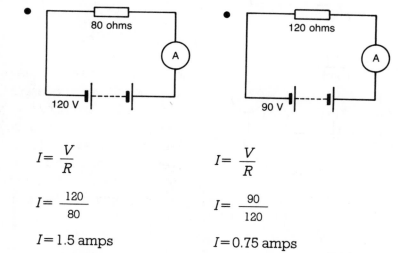

$I = \dfrac{V}{R}$ $I = \dfrac{V}{R}$

$I = \dfrac{120}{80}$ $I = \dfrac{90}{120}$

$I = 1.5$ amps $I = 0.75$ amps

Now look at the formula again.

$$I = \frac{V}{R} \left(\text{amps} = \frac{\text{volts}}{\text{ohms}} \right)$$

We may rearrange this two more ways.

$$R = \frac{V}{I} \left(\text{ohms} = \frac{\text{volts}}{\text{amps}} \right)$$

and $V = IR$ (volts = amps × ohms)

These three versions of the formula let us work out any one of volts, amps or ohms if we know the other two.

Examples

- What is the value of the resistor?

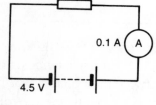

$R = \dfrac{V}{I} = \dfrac{4.5}{0.1} = 45$ ohms

- What is the battery voltage?

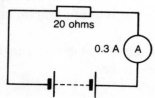

$V = IR = 0.3 \times 20 = 6$ volts

There is no need to learn all three version of the formula. You may simply learn this triangle.

To find the current look in the triangle. V is over the top of R so

$$I = \frac{V}{R}$$

If you want to calculate a resistance and you know volts and amps, then V is over I, so

$$R = \frac{V}{I}$$

I is next to R in the triangle, so just multiply them to find V.

$$V = IR$$

So the triangle is the only thing you have to learn and a calculator will do the sums for you! This is the most useful formula in electronics. You can work out lots of things by using it.

Questions

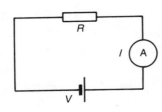

1. R is 20 ohms, V is 30 volts, what is I?
2. I is 0.2 amps, R is 50 ohms, what is V?
3. I is 0.4 amps, V is 20 volts, what is R?
4. R is 18 ohms, I is 500 mA what is V?

In question (4) the current was given as 500 mA, but the formula is $V = IR$ (volts = amps × ohms). The formula says amps, it does *not* say milliamps. If your answer was 1000 times too big, this might be why! Always *check* that you use the unit the formula needs.

All this was realised by Georg Simon Ohm in 1827. He worked out *Ohm's law* which you can find in physics books.

Task

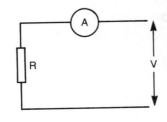

Make up the circuit shown. Measure the current through different resistors and calculate what the current should be. Does it correspond with your meter reading? If it doesn't, check the actual voltage of your battery with the multimeter set to read volts.

Two precautions

- Only use resistors larger than 100 ohms.
- Don't forget that your meter will probably read milliamps. The formula uses amps.

Task

Copy this summary into your notebook, filling in any blanks.

Resistance

Resistance is measured in .... The larger the ... the less is the ... which flows. The larger the ... across a resistor the ... is the ... which flows.

$$I = \frac{V}{R} \qquad R = \frac{V}{I} \qquad V = IR$$

Resistors

Ways of labelling resistors

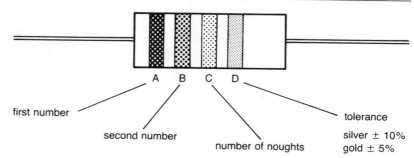

first number
second number
number of noughts
tolerance
silver ± 10%
gold ± 5%

Resistors have coloured bands on them. The colours indicate the *value* in ohms. Resistors always have three coloured bands; they often have four. The colours used are

black	brown	red	orange	yellow	green	blue	violet	grey	white
0	1	2	3	4	5	6	7	8	9

You start reading the colours at the end *away* from the gold or silver.

If the resistor was   red   yellow   orange   silver
                                 ↓   ↓   ↓   ↓
it would have a value of   2   4   000   10%

Note The third colour is the *number of noughts*. So the resistor above is *not* 243 ohms, it is 24 000 ohms.

The *tolerance* tells us the accuracy to which the resistor has been made. Few electronic circuits need very accurate resistors. ±1% tolerance resistors are quite expensive. ±10% is good enough for most purposes.

Another way of marking a resistor is to write on it.

       100 ohms is written   100R
       820 ohms is written   820R
    1000 ohms is written   1K   K means 1000
    6800 ohms is written   6K8
1 000 000 ohms is written   1M   M means 1 000 000
1 200 000 ohms is written   1M2
      1.2 ohms is written   1R2

This is a variable resistor

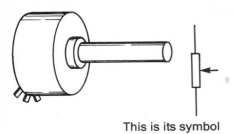

This is its symbol

Sometimes the sign Ω is used to mean ohms. 100Ω is 100 ohms.

Resistors are made in only a limited number of sizes. One series, called the *E12 series*, has in it 10R, 12R, 15R, 18R, 22R, 27R, 33R, 39R, 47R, 56R, 68R, 82R, 100R, 120R, 150R and so on up to 10M.

There are also *variable* resistors. You use one of these each time you adjust the volume on a radio.

Temperature and resistance

Don't think that the only thing in a circuit which has resistance is a resistor. Other components all have resistance, some of them have very large resistance. Even the wiring has some resistance. The dry joint on page 11 has a high resistance. This is accidental and causes circuits to fail. Also, the resistance of metal wires and resistors varies with temperature. The resistance of these parts of a circuit get higher as the temperature goes up.

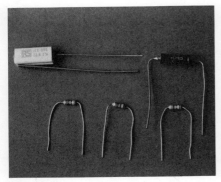
Resistors

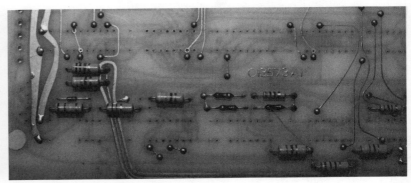
Resistors in a circuit

Connecting two resistors

What happens if you have two resistors connected together? You will remember there are two ways of doing this, in *series* and in *parallel*.

Series
Here the current which flows is reduced by each resistor.

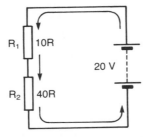

The formula is
$$I = \frac{V}{R}$$

There are two resistors in series, so you just add them.

$$I = \frac{V}{R_1 + R_2}$$

So in this case

$$I = \frac{20}{10+40}$$

$$I = \frac{20}{50}$$

$I = 0.4$ amps

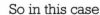

0.4 amps will flow in each resistor.

Parallel

Here *part* of the current goes through R_3 and part through R_4.

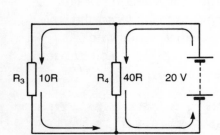

The formula is

$$I = \frac{V}{R}$$

Because there are two routes for the current you may work it out for each in turn.

$$I = \frac{V}{R_3} \qquad I = \frac{V}{R_4}$$

$$I = \frac{20}{10} \qquad I = \frac{20}{40}$$

$$= 2 \text{ amps} \qquad = 0.5 \text{ amps}$$

So the total current is $2 + 0.5 = 2.5$ amps.

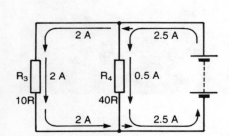

The voltages have also to be worked out differently.

Series

We have already seen how to do this using

$$V_{R_1} = \frac{R_1}{R_1 + R_2} \times V_{\text{baH}}$$

$$V_{R_1} = \frac{10}{50} \times 20$$

$$= 4 \text{ volts}$$

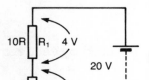

But there is another way using the 'Ohms law' calculation from page 22.

We have seen that the current is 0.4 amps in both R_1 and R_2. We can use $V = IR$ to find the voltages.

$$V_{R_1} = 0.4 \times 10$$
$$= 4 \text{ volts}$$

What is V_{R_2}? Calculate it.

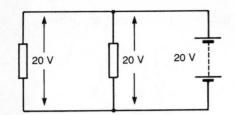

Parallel
The voltage is exactly the same across each resistor.

Questions

1. Write the following in full (e.g. 6K8 is 6800 ohms).
   a) 3K9   b) 39K   c) 47R
   d) 2M2   e) 8K2

2. Write the following in the brief way (e.g. 6K8).
   a) 2700 ohms   b) 1 000 000 ohms   c) 1 200 000 ohms
   d) 5600 ohms   e) 18 ohms   f) 330 000 ohms

3. Write down the colours you would find on the following resistors.
   a) 15K   b) 1K5   c) 150R
   d) 1M   e) 330K   f) 12R

4. What value would resistors with the following colours have?
   a) red red orange.   b) green blue brown
   c) brown red black   d) orange orange orange
   e) red violet yellow   f) orange violet red

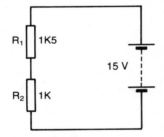

5. What current would flow in R_1?
   a) 0.01 A   b) 100 mA
   c) 6 A   d) 6 mA

6. In the circuit in question (5), what current would flow in R_2?
   a) 0.015 A   b) 66 mA
   c) 0.015 mA   d) 6 mA

7. In the circuit in question (5), what would be the voltage across R_1?
   a) 6 V   b) 12 V   c) 9 V   d) 15 V

8. Which of the answers in question (5) would be correct for the voltage across R_2?

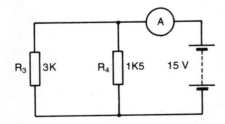

9. The current in R_4 is
   a) 10 mA   b) 4.28 A
   c) 5 mA   d) 0.2 A

10. In the circuit in question (9), the current in R_3 is
    a) 10 mA   b) 0.003 A
    c) 5 mA   d) 5 A

11 In the circuit in question (9), the total current passing through meter A is
   a) 10 mA   b) 0.015 A
   c) 5 mA   d) 1.5 A
12 In the circuit in question (9), the voltage across R_3 is
   a) 1.5 V   b) 10 V
   c) 5 V   d) 15 V
13 In the circuit in question (9), the voltage across R_4 is
   a) 1.5 V   b) 10 V
   c) 5 V   d) 15 V

What is Electricity?

Current electricity

Everything is made of atoms which are very small indeed. There are about 92 different sorts of atom found in nature. Atoms have two main parts. These are the *nucleus* in the centre and a cloud of *electrons* moving round the outside. Each electron has a *negative* electrical charge. These are normally balanced by the same number of *positive* charges on the nucleus.

Unlike charges *attract* each other. Like charges *repel* each other.

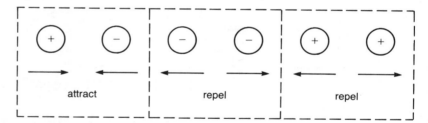

We may imagine that the simplest atom, hydrogen, might look like this.

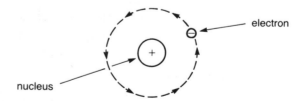

The next most complicated atom is helium. It has two electrons and two positive charges in the nucleus.

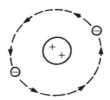

The next on the list is lithium. How many electrons does it have? The electrons which go round the nucleus are held there because unlike charges attract. In metal atoms like copper, the electron furthest out is only loosely held. It is called a *free electron*.

Imagine a line of copper atoms.

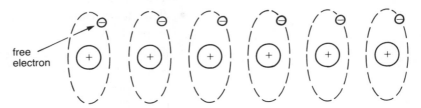

Now imagine we connect this line of copper atoms to a battery. The negative terminal of the battery *repels* the free electrons, and the positive terminal *attracts* them. It is this *movement* of free electrons along wires that is an electric current. This is *current electricity*.

Static electricity

You can charge up *insulators* as the following experiments show.

Tasks

1. Rub a plastic pen or comb on your sleeve. It will then pick up very small pieces of paper. This is because you have moved some electrons from your sleeve onto the plastic. So the plastic is *negatively* charged. Because you have removed electrons from your sleeve that is now left with a *positive* charge.
2. Now hold the charged pen near a very thin stream of water from a tap. You should find that it attracts the water.

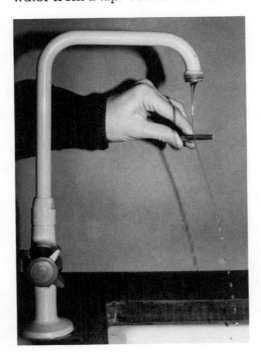

These effects are due to charges which do not move. This is called *static electricity*. The voltages may be very high but the currents are small.

Notes
- Try several plastic pens to find one which works well. Plastics vary.
- The description of atoms used here is very greatly simplified.
- Electricity was discovered very many years before electrons. There was a convention that electricity flowed from positive to negative. This is now called *conventional flow*. But we know electrons go from negative to positive. This is called *electron flow*.

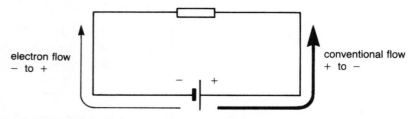

Task

Copy this summary into your notebook, filling in the blanks.

Electricity
Electric currents are a flow of ... from ... to plus. Like charges .... Unlike charges .... Current which ... along conductors is ... electricity and electricity which stands still on insulators is ... electricity.

Questions

1. Atoms
   a) consist of a positive nucleus and electrons
   b) consist of a positive nucleus and negative electrons
   c) consist of a negative nucleus and positive electrons
   d) have different types of charge according to the sort they are.
2. Choose the correct statement.
   a) Like charges may attract or repel
   b) Unlike charges never attract
   c) Like charges repel, unlike charges attract
   d) Unlike charges repel, like charges attract.
3. Electrical flow
   a) is electrons flowing from negative to positive
   b) is electrons flowing from positive to negative
   c) is a nucleus flowing from negative to positive
   d) is always called static.
4. Conductors
   a) never have free electrons
   b) always have free electrons, just like insulators
   c) may be made of the same material as insulators
   d) have free electrons.

The Cathode Ray Oscilloscope

The cathode ray tube

The oscilloscope lets you see electrical voltages and the way they may change very quickly. The most important part is the *cathode ray tube*.

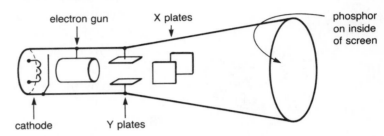

Electrons are produced by the cathode. They are accelerated by the gun and strike the screen. The screen is coated with a *phosphor* which glows when struck by electrons, so a spot of light is produced. You may connect things to the Y-plates through terminals on the front of the oscilloscope. What would happen to the electron beam if the top Y-plate was connected to positive of a battery and the bottom Y-plate to negative? *Remember*: electrons have a negative charge. Like charges repel. Unlike charges attract.

You should have worked out that the beam will go up and the spot of light will move higher on the screen.

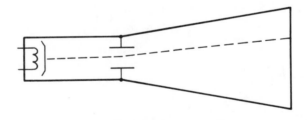

The oscilloscope has several quite complicated circuits to produce different voltages, amplify and move the beam.

Direct and alternating current

You will need to know how to operate an oscilloscope. In the following activities, details are given for an excellent simple oscilloscope. Its controls are shown at the bottom of page 33. Some oscilloscopes have many more.

Tasks

1. Plug in and switch on.
2. Turn the *time base* off. In the oscilloscope shown in the photograph, the 'ms/div' switch is turned fully anticlockwise.
3. Turn the *AC/DC* switch to DC.
4. Turn up the brilliance and adjust the *X shift* and *Y shift* until the spot of light is in the centre of the screen.
5. Reduce the brilliance and adjust the *focus* control to get a sharp spot of light.

Now connect a battery to the *Y input* terminals. The spot may move up or down. Work out which terminal is connected to the top Y-plate. (*Remember*: like charges repel.) If the spot hardly moves, or goes right off the screen, then adjust the *volts/div* control. This changes the *sensitivity* of the oscilloscope. 'Volts/div' means *volts per division*.

If the switch is set to 1 and the spot goes up three divisions on the screen, then what is the voltage connected to the Y-plates?

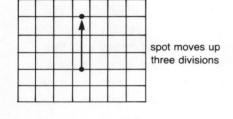

spot moves up three divisions

Use the oscilloscope to find the voltage of your battery.

Now switch the time base on. To do this on the oscilloscope shown, turn the ms/div control to 100. The spot will move from left to right across the screen. Turn the time base one click further and the spot will go ten times faster. Turning the *variable* control gives fine adjustment of the speed.

Connect up your battery again. You will have a line across the screen at a height which depends on voltage. This is *direct current* or *DC*. The voltage of the battery does not change from moment to moment and you can see this.

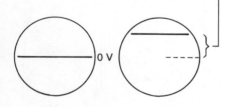

battery not connected

battery connected

height of line depends on voltage

Switch the time base off again. Centre the spot of light, using the X shift. Connect the Y input to the AC low voltage terminals of a power supply unit. **Do not connect to the high voltage mains – they are dangerous**. The trace should appear like this.

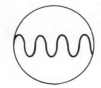

What is the voltage doing? It is changing rapidly from positive to negative.

Switch on the time base again and adjust it to get a trace which looks like this.

This is *alternating current* or *AC*. The electric power supply of the mains changes up and down (or *alternates*) like this 50 times each second. We say it has a *frequency of 50 hertz* (50 Hz for short).

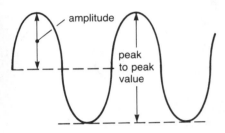

We may use an oscilloscope to measure the *amplitude period* and *peak to peak* voltage of AC.

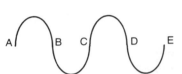

The *period* is the time taken for one complete cycle. This is the time needed for the voltage to change from A to C or B to D in the diagram.

Can you see another interval of exactly one period?

Question

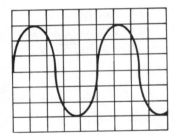

This diagram shows AC on an oscilloscope which is set to 2 volts/div and 1 ms/div (1 ms means 1 millisecond). What is a) the amplitude, b) the period in milliseconds, c) the peak to peak voltage?

You can make AC by moving a magnet in and out of a coil of wire. Connect the coil to the Y input to see the results.

How does its frequency compare with the 50 Hz of the power supply unit?

Producing AC with a magnet and coil

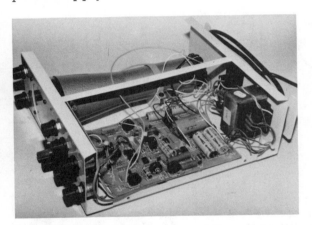

Inside the oscilloscope

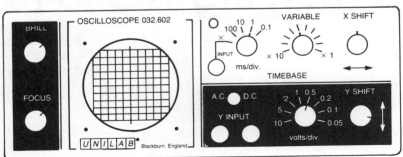

The oscilloscope controls

Task

Copy the diagram of the cathode ray tube into your notebook.

Questions

1. Each diagram shows a trace on an oscilloscope screen. Match each diagram with one of these descriptions.

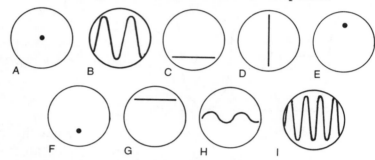

a) time base on, top Y plate connected to +
b) time base off, top Y plate connected to −
c) time base off, top Y plate connected to +
d) time base on, top Y plate connected to −
e) time base off, Y plates connected to AC
f) time base off, Y plates not connected
g) time base on, Y input connected to 6 volts, 50 Hz AC
h) time base on, Y input connected to 6 volts, 100 Hz AC
i) time base on, Y input connected to 2 volts, 50 Hz AC

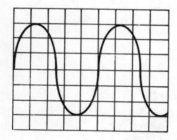

2. The diagram shows AC on an oscilloscope screen. The sensitivity is set to 1 volt/div. The amplitude is
   a) 3 V   b) 6 V   c) 9 V   d) 12 V

3. Which of the answers to question (2) is the peak to peak voltage?

4. The time base is set to 10 ms/div. The period is
   a) 2.5 ms   b) 5 ms   c) 25 ms   d) 50 ms

Rectifying AC to Produce DC

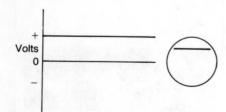

Most electronic devices need a DC supply but mains electricity is AC. DC flows in one direction only. You can label the wires + and −.

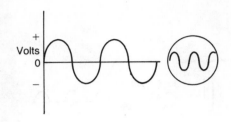

AC flows to and fro. The + and − wires change over 50 times each second. You cannot label the wires + or − because they are changing all the time.

To turn AC into DC we use a *diode* as a *rectifier*. The rectifier diode may pass much more current than an LED and it does not light up like an LED. *Diodes* pass current in *one direction only*.

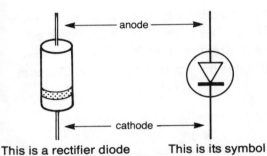  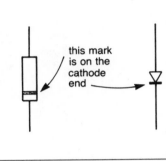

This is a rectifier diode   This is its symbol   or

Task

You can show how a diode works with a torch bulb and a battery. Connect these circuits.

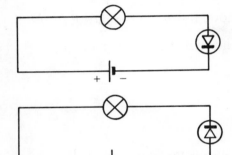

- The bulb lights. The diode is *forward biased* and it conducts.

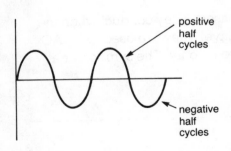

- The bulb does not light. The diode is *reverse biased* and it does not conduct.

What do you think would happen if you reversed the battery connections? Try it to see.

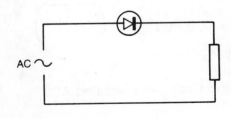

AC reverses its direction many times each second.

If we connect AC to a diode only the *positive half cycles* are conducted. Here the voltage across the resistor will be like this. ⌒⌒⌒⌒

35

Tasks

1. Connect up this circuit.

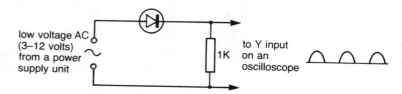

Adjust the time base and sensitivity and you should obtain a trace like that shown above. This is *half wave rectified* AC. It is flowing in one direction only, like DC.

2. A better rectifier circuit is the *full wave rectifier*. Connect this circuit.

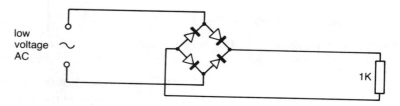

Connect an oscilloscope across the 1K resistor. You should get a trace which shows full wave rectification.

Note When wires cross and *join*, they are drawn like this.

If they do *not join*, they are drawn like this.

3. Copy this summary into your notebook, filling in the blanks.

Diodes

Diodes conduct when . . . biased. They conduct when the . . . end is connected to positive. They are used to . . . AC because they conduct in one . . . only. The symbol for a diode is . . . .

Questions

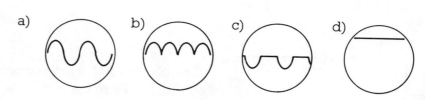

1. Which of the traces above shows half wave rectified AC?
2. Which trace shows AC?

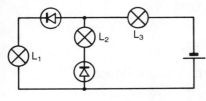

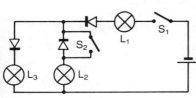

3. Which statement is true for this circuit?
   a) Only L_3 will light.
   b) Only L_2 will light.
   c) Only L_2 and L_3 will light.
   d) Only L_1 will light.
4. In the circuit shown, S_1 is closed (switched on), S_2 is open. Which lamps will light?
   a) L_1, L_2 and L_3
   b) L_1 and L_2 only
   c) L_1 and L_3 only
   d) L_3 only
5. In the circuit of question (4), both S_1 and L_2 are now closed. Which of the answers to question (4) is now correct?

A useful hint

It is easy to work out whether a diode will conduct from a circuit diagram. Imagine the symbol is like an arrow.

Then the diode will conduct if the arrow points in the direction of the *conventional* (+ to −) flow.

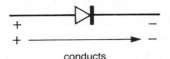
conducts

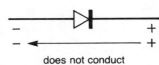
does not conduct

Capacitors

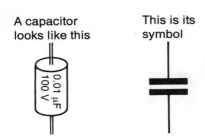

A capacitor looks like this   This is its symbol

Capacitors are components which will store a small amount of electricity. They will pass varying currents (AC) but will block steady currents (DC).

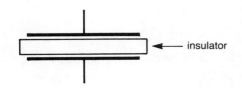
← insulator

Inside a capacitor there are two metal plates separated by an insulator. Usually the metal plates are made of foil and the whole thing is rolled up to make the capacitor small in size. They may be connected either way round.

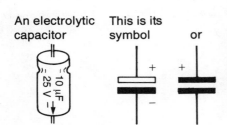

An electrolytic capacitor   This is its symbol   or

Capacitors are made with different values. The values are measured in *microfarads*. Microfarad is abbreviated to μF. Large capacitors, over about $10\,\mu F$, are usually made in a different way to small ones. They have a chemical as an insulator. They are called *electrolytic* capacitors.

Electrolytic capacitors *must* be connected the right way round. They do not work if wrongly connected.

37

Tasks

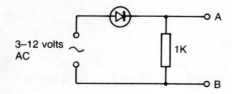

1 Connect up a half wave rectifier.

Connect A and B to the Y input of an oscilloscope. The trace should look like this.

This half wave rectified current is too uneven to use for such things as a radio. But a capacitor will smooth the current.

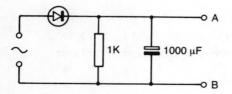

2 Connect this circuit.
The trace should look like this.

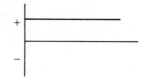

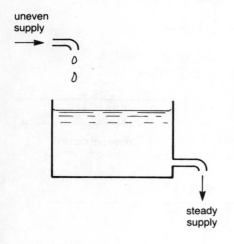

The capacitor has *smoothed* the current. It does this because it charges up to the supply voltage. It is rather like an uneven jerky water supply being turned into a steady one by a small storage tank.

Task

Charging and discharging a capacitor

Arrange a 1000 μF capacitor and a 2.5 volt torch bulb in series like this.

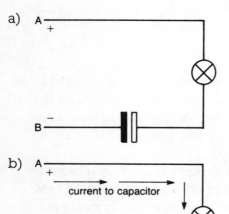

Touch ends A and B to a 5 or 6 volt DC supply. The bulb will flash for a moment, then go out. This is because current flows through the bulb to charge the capacitor. The insulator stops it flowing continuously, so the bulb just flashes. The capacitor does not pass DC continuously.

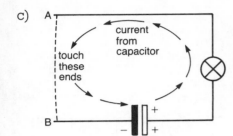

Now touch the two ends A and B together. The bulb will again flash. This is because the capacitor now discharges through the bulb. You can repeat these steps but you will *always* have to charge then discharge, then charge again, and so on, in that sequence. Why is this?

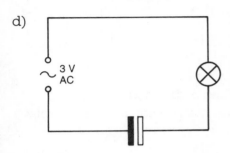

Now connect the same circuit to about 3 volts AC. The bulb will stay alight. AC is changing direction all the time at 50 Hz. The capacitor is being charged and discharged 50 times each second, so the bulb lights.

Markings on capacitors

The *microfarad* is *one millionth* of a farad or 0.000 001 farad. Even the microfarad is too big for many purposes, so we also have the *nanofarad* (nF), 0.001 of a microfarad, and *picofarad* (pF) 0.000 001 of a microfarad.

Capacitors may be made small in size but hold a lot of charge if the insulator is thin, *but* the thinner the insulation is then the lower is the voltage which it will stand. So capacitors are made in the same value for several *different* voltages. For example, you may obtain a 47 μF capacitor for use on 10 volts, 25 volts or 63 volts. Which would you expect to be largest? Why? The working voltage is often marked on the capacitor, e.g. 47 μF, 25 V.

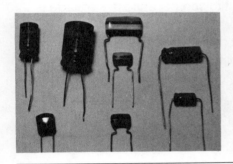

Tasks

1 Try charging and discharging the 1000 μF capacitor through an LED and resistor. The low current needed by the LED makes the effect last longer than with a bulb.

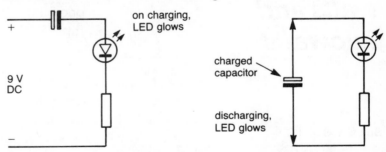

2 Copy this summary into your notebook, filling in the blanks.

Capacitors

Capacitors will block ...... but allow ...... to flow through them. They hold a small e...c charge. This effect is used for s... half wave rectified AC. E....... must be connected the right way round. The symbol for an electrolytic capacitor is ... and for an ordinary capacitor is ....

Questions

1. Capacitors will
   a) pass AC and block DC
   b) pass DC and block AC
   c) pass AC and DC
   d) pass neither AC or DC.

2.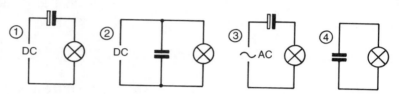

   The bulbs will light in
   a) 3 only b) 2 only c) 2 and 3 d) 1 and 4

3. The bulb in question (2) will flash on connection but not run continuously in circuit
   a) 1 b) 2 c) 3 d) 4

4. If you charge a capacitor through a bulb, the bulb flashes. If you recharge again, you will not see the bulb flash because
   a) the capacitor is already charged, so no more current can flow into it unless it is first discharged
   b) the bulb flash is too fast to see because the voltage is higher
   c) electrolytic capacitors do not charge quickly
   d) recharging it takes current out of it.

Electrical Power

Milliwatts, watts and kilowatts

In one bulb, 2 amps flow.

Another bulb is connected to 6 volts.

Neither of these statements tells us very much about either circuit. Knowing the voltage or the current on its own tells us little. But if we know *both* we can find out a great deal more.

The following table gives some information about car light bulbs.

Bulbs A, B and C

Bulb	Supply voltage V	Current which flows I
A	12 V	0.5 A
B	12 V	2 A
C	12 V	4 A

Which bulb would you expect to give most light? C would be brightest. In everyday language, C is the *most powerful*.

Now here are two more bulbs.

Bulb	V	I
D	250 V	1 A
E	100 V	2 A

Which is the most powerful now, C, D or E? It is now not so obvious as *both I* and *V* vary.

To find out which is brightest you have to calculate the *power*. The unit of power is the *watt* (W). The calculation is

power = voltage × current

In units   watts = volts × amps.
In symbols $P = VI$.

Example

What power is the bulb which passes 2 amps at 12 volts?
$P = VI = 12 \times 2 = 24$ watts

Now work out the power of bulb A, C, D and E above. The answers you should have obtained are: A, 6 watts; C, 48 watts; D, 250 watts; E, 200 watts.

Electrical *power* does not depend on current flowing alone, nor does it depend only on the EMF driving the current. It depends on *both* current *and* voltage, multiplied together.

In circuits where the power is very small *milliwatts* (mW) are used.

1 milliwatt = 0.001 watts
1000 milliwatts = 1 watt

When amounts of power are large we use *kilowatts* (kW).

1 kilowatt = 1000 watts

Size of resistors

Look at the diagram of resistors. They are drawn full size and all are 1000 ohm (1K) resistors. Why are the sizes different?

a)   b)   c)

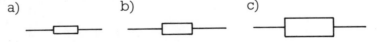

1000 Ω resistors

It is because they handle different amounts of power; a) is 0.25 W, b) is 0.5 W and c) is 2.5 W.

Low power resistors are smaller and cheaper than high power ones. It is very important to use resistors, and some other components, with big enough power ratings. If you do not they will overheat and fail. *Never* replace a large dimension resistor with a smaller one. Even if they have the same value in ohms, they handle different powers.

Questions

Calculate the power of the following. Write the answers in kilowatts *or* milliwatts as well as watts.

1. bulb, 24 V, 5 A
2. motor, 12 V, 3 A
3. drill, 250 V, 1.5 A
4. heater, 250 V, 6 A
5. kettle, 240 V, 8 A
6. calculator, 3 V, 0.001 A

Example

We can use the $P = VI$ formula to calculate currents. A TV set has a label at the back '250 V, 60 W'. What current flows in the mains lead?

$P = V \times I$ *Write the formula.*

$60 = 250 \times I$ *Substitute the figures you know.*

$\dfrac{60}{250} = I$ *Do the calculation.*

$= 0.25$ amps   *Put in the unit.*

Questions

1. What current flows in a 6 V, 6 W lamp?
2. What is the largest current a 2 watt resistor should pass if connected to 4 volts?

Task

Copy this summary into your notebook, filling in the blanks.

Power
To calculate electrical power you ... volts by ....
1000 watts = ... 0.001 watts = ...
$P = I$ ....

Questions

1. Copy and complete the table, filling in the blanks.

	EMF	Current	Power
a)	24 V	3 A	?
b)	5 V	0.1 A	?
c)	9 V	0.06 A	?
d)	2 V	?	0.5 W
e)	5 V	?	2 W

2. Write in milliwatts a) 2 W, b) 0.2 W, c) 0.02 W.
3. Write in kilowatts a) 2500 W, b) 250 W, c) 5000 W.
4. Write in watts a) 1500 mW, b) 500 mW, c) 1.2 kW, d) 0.5 kW.

Transformers

Changing AC voltages and currents

Transformers are devices which step alternating voltages up or down and at the same time change currents. They consist of two coils of wire wound onto a single soft core. The photographs show several types.

These are transformers

This is the symbol for a transformer

primary coil   secondary coil

1 V AC   2 V AC

10 turns   20 turns

When AC is connected to one coil, called the *primary*, then an AC supply is obtained from the other coil, called the *secondary*.

If the secondary coil has twice as many turns as the primary then the secondary voltage is twice the primary voltage. If the secondary has ten times as many turns as the primary then the secondary voltage is ten times the primary, and so on. Transformers will step voltages down if the secondary has fewer turns than the primary coil.

Task

Wind a coil of ten turns onto a soft iron core. Then wind a 20 turn coil onto another core and put the two cores together.

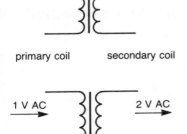

10 turns   20 turns

soft iron cores

Now connect the 10 turn coil to 1 or 2 volts AC. Connect either a multimeter set to read AC volts *or* an oscilloscope Y input to the 20 turn coil.

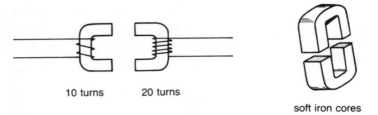

1 V AC   primary 10 turns   secondary 20 turns

1 V AC   primary   secondary   to oscilloscope

You should find that the secondary voltage is about twice the primary voltage.

43

Now change the transformer connections over. Make the 20 turn coil the primary by connecting it to the AC supply and connect your new secondary to the measuring device. Should the new secondary voltage be greater or less than before? It should be about half the supply voltage.

Try making a transformer which steps the voltage up five times. With ten turns on the primary, how many turns will you need on the secondary?

There is a formula for working out the voltage.

$$\frac{\text{secondary volts}}{\text{primary volts}} = \frac{\text{secondary turns}}{\text{primary turns}}$$

In symbols $\quad \dfrac{V_s}{V_p} = \dfrac{T_s}{T_p}$

Example

A primary coil of 20 turns is connected to a 15 volt supply. What voltage will be available from a 40 turn secondary?

$$\frac{V_s}{V_p} = \frac{T_s}{T_p}$$

$$\frac{V_s}{15} = \frac{40}{20}$$

$$20 \times V_s = 15 \times 40$$

$$V_s = \frac{15 \times 40}{20}$$

$$V_s = 30\,\text{V}$$

In electronics you do not get something for nothing. When the secondary voltage is stepped up the current which is available is reduced. Look at this transformer.

1 V AC   2 V AC
1 A   0.5 A

The secondary voltage is twice the primary voltage but the secondary current is half the primary current. Can you recognise the quantity which is the *same* for both primary and secondary flows of electricity?

Hint Look back to page 41.

It is the *power* which is the same in each ($P=VI$).
primary watts $= VI = 1 \times 1 = 1$ watt
secondary watts $= VI = 2 \times \tfrac{1}{2} = 1$ watt

You may change voltage *or* current up or down as you wish, but the *power* stays the same. In fact the output power is slightly less than the input power because the transformer is not 100% efficient. However they do so well that for a lot of purposes we may ignore the slight losses.

Tasks

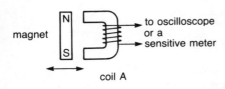

coil A

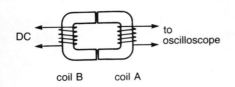

coil B   coil A

1 How does a transformer work?
Connect up a 30 turn coil on a core to an oscilloscope. Move a magnet to and from the coil.

You should obtain a trace which shows that a rather untidy looking AC is made. This is the way a dynamo works – as invented by Michael Faraday.

Now connect up a second coil to low voltage *DC* or a battery. Move B to and fro. Current is again made in coil A. Coil B with a steady current is an *electromagnet*. If you now connect coil B to AC we are back with the transformer again. The AC gives the effect of moving coil B to and fro 50 times each second.

2 Copy this summary into your notebook, filling in the blanks.

Transformers
A transformer steps voltage up or ..... It has a primary c... and a... coil. If it steps voltage... then the... goes down and vice versa. Transformers will only work on... current.

$$\frac{\text{secondary volts}}{\text{primary volts}} = \frac{?}{?}$$

Questions

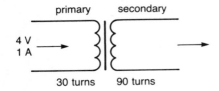

1 The secondary voltage is
   a) 1 V   b) 4 V   c) 12 V   d) 36 V
2 In question (1), the power in the primary coil is
   a) 1 W   b) 4 W   c) 12 W   d) 5 W
3 In question (1), the secondary current is
   a) $\tfrac{1}{3}$ amp   b) 1 amp   c) 3 amps   d) $\tfrac{1}{4}$ amp

More about Diodes

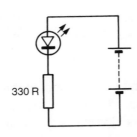

This is the circuit we used on page 7 to make an LED work.

An LED will pass a maximum current of about 25 mA (0.025 A). More than this is very likely to damage the LED, but the resistor R limits the size of the current. We shall now see how to calculate the resistor size.

Task

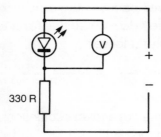

Set up this circuit.
Gradually increase the voltage in steps from 1 volt to about 6 volts. When does the LED light?

You will find that the LED does not work until there is a voltage of about 2 volts across the LED. There has to be a voltage across it of about 2 volts before it starts conducting. Whenever you connect a meter across a working LED you will always find this 2 volt different. It is called the *forward voltage*.

To find the size of resistor R we just subtract the forward voltage from the supply voltage and divide that by the current (less than 0.025 mA).

$$\text{resistance} = \frac{\text{supply volts} - \text{forward volts}}{\text{current}}$$

In symbols $\quad R = \dfrac{V_S - V_F}{I}$

This looks frightening but it is really just a slightly more complicated version of Ohm's law, $\quad R = \dfrac{V}{I}$

Example
An LED is connected to 8 V. What resistor is needed if the current is to be 20 mA (0.02 A)?

$$R = \frac{V_S - V_F}{I}$$

$$R = \frac{8 - 2}{0.02}$$

$$R = \frac{6}{0.02}$$

$$= 300 \text{ ohms}$$

The nearest preferred values are 270 Ω and 330 Ω. Which would you choose? *Remember*: you must *limit* the current – stop it rising above 20 mA.

Diodes always have this forward voltage across them when they conduct. For an ordinary diode made of silicon it starts conducting only when the forward voltage is about 0.7 volts.

Zener diodes

Zener diodes are used in a completely different way from other diodes. They control a supply voltage *very* precisely. You might think that a mains power supply voltage is absolutely steady, but this is not the case unless we take steps to control it. For example, a radio playing loudly will draw more current than one playing quietly. This changing current load may affect the voltage supplied. Imagine a car travelling along a hilly road. The *load* on the engine varies. The engine speed goes up and down with the slope. The *load* of an electronic circuit on its power supply may vary too. A radio playing music which is changing in volume is making rapid changes in demand on its power supply. Zener diodes enable us to *stabilise* the power supply.

A zener diode looks like an ordinary diode. It will have a voltage marked on it, e.g. 4.7 V, 6.2 V, etc. This is called the *zener* voltage, V_z.

The symbol for a zener diode

Task

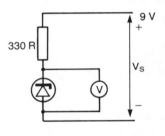

Set up this circuit.

You will find that the voltage across the zener diode is the same as that marked on the zener diode itself. Test this by trying other diodes, e.g. 6.2 V, 6.8 V, etc.

Now reduce the supply voltage V_S to less than the zener voltage. What does the meter show now?

Look at the circuit carefully. What do you notice about the arrangement? You should see that the zener diode is *reverse biased*. It does not conduct until its zener voltage is reached. Then it does conduct very well.

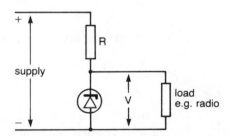

In practice the diode is used like this. The voltage V supplied to the load will equal the voltage marked on the zener diode. The only condition is that the supply voltage has to be at least as large as the zener voltage.

The resistor R is there to prevent too much current passing through the diode. Zener diodes have *power* ratings as well as voltage ratings. To find the resistance of resistor R we use the power rating to find the current, then Ohm's law.

Example

An 8 volt zener diode has a maximum power of 400 mW (0.4 watts) and is connected to a 12 V supply.
What resistor should be used in series with it?

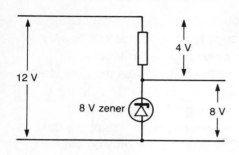

Step 1

Find the current flowing. Watts = volts × amps

$$0.4 = 8 \times \text{amps}$$
$$\frac{0.4}{8} = \text{amps}$$
$$= 0.05 \text{ amps}$$

Step 2

The power supply is 12 V and the zener diode is an 8 V type; so 4 V appears across the resistor (12−8=4 V). Find the resistor which passes 0.05 A at 4 V.

$$R = \frac{V}{I}$$
$$= \frac{4}{0.05}$$
$$= 80 \text{ ohms}$$

Question

A 10 V zener diode has a maximum power of 0.5 W and is connected to a 12 V supply. What resistor should be used in series with it?

Task

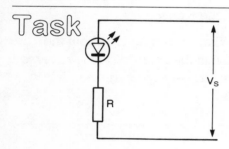

Copy this summary into your notebook, filling in the blanks.

More about diodes

LEDs have a ... voltage drop across them of about ... volts when they are c...g. They do not start to conduct until this ... ... is ... volts.

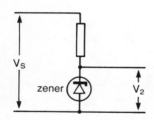

To find the resistance R of resistor R use the formula.

$$R = \frac{V_S - V_F}{I}$$

Zener diodes are used ... biased to control supply voltages. The output voltage is the zener voltage V_z provided that the input voltage is ... than V_z.

Questions

1. LEDs have a forward voltage of about
   a) 0.2 V   b) 0.7 V   c) 2.0 V   d) 2.6 V
2. Which of the answers to question (1) is correct for a silicon diode?
3. The LED will pass a maximum of 0.02 A. R should have a value of at least
   a) 500 ohms   b) 450 ohms   c) 400 ohms   d) 350 ohms

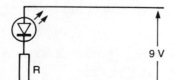

4 Z is 1 W, 5 V zener diode. The current it will pass is
   a) 0.1 A   b) 0.2 A   c) 0.3 A   d) 1.0 A
5 The resistor R should have a value of at least
   a) 20 ohms   b) 50 ohms   c) 16 ohms   d) 35 ohms

6 The output voltage will be
   a) 0 volts   b) 8 volts   c) 7 volts   d) 2 volts

Transistors

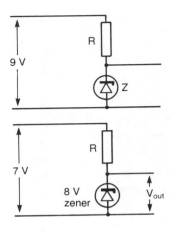

Transistors use small electric currents to control much larger currents. They are used in radios, televisions, calculators and computers.

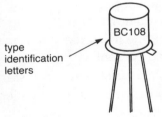

This is a transistor

This is its symbol

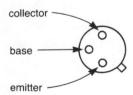

The three connections on a transistor

There are three connections on a transistor. These are the *base*, the *emitter* and the *collector*. It is important to identify the leads so that you connect them correctly. Look at the underneath of the type BC108 transistor. You will see three leads. You can see which lead is which by its position.

Task

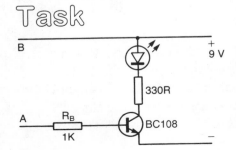

Connect up this circuit. Do not connect A to anything yet. The LED will not light. No current flows through the emitter and collector. No current flows through the base and 1K resistor because they are not connected.

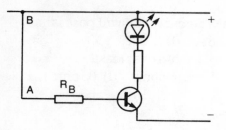

Now connect A to the positive supply as shown. You should find that the LED now lights. This is because a small current of about 1 mA (0.001 A) flows through the base and emitter. When this base-emitter current flows, then a much larger current flows through the collector and emitter.

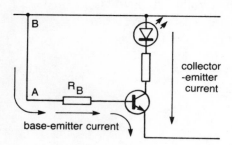

So the larger collector-emitter current is controlled by the very much smaller base-emitter current. The resistor R_B is there to prevent too much current flowing through the base. If you leave out R_B a much bigger current will flow through the base-emitter. This will cause such a large current to flow through the collector-emitter circuit that the transistor will be destroyed. *Moral*: never leave out some means of limiting the base current.

Another way of thinking about a transistor is like this. Imagine it is a resistor whose resistance is controlled by the base current.

The resistance is *transferred* from base to collector. The device was originally called a *transfer resistor* – later shortened to *transistor*.

Transistors are made in thousands of different types. They handle different currents, work at different frequencies and may produce various collector currents for given changes in base current. Here are some examples.

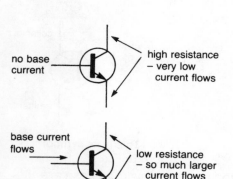

Type	Maximum collector current	Maximum power	h_{FE}
BC108	100 mA	0.3 W	125
BFY51	1 A	0.8 W	40

Why don't we use the BFY51 all the time since it seems more robust than a BC108? The answer is that the BC108 is much cheaper and it has a higher h_{FE}. The h_{FE} shows how the collector current changes with base current. High h_{FE} numbers show that collector current changes a lot for a small change in base current.

Task

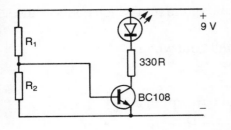

In use, transistors are controlled by *two* resistors in series. This is the potential divider circuit from page 18.

Connect this circuit. Use an 8200 ohm resistor for R, and try the following for R_2. First 200R, second 470R, third 680R.

You should find that the LED in the collector does not light with the first or second resistor but it does with the third one. This is because the voltage between base and emitter has to be at least 0.7 volts to drive enough current into the base to turn the transistor on. This base-emitter voltage of about 0.7 volts is the same for all transistors made of silicon.

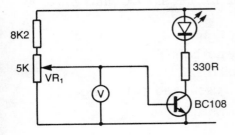

Now connect up this circuit. VR_1 is a variable resistor. Adjust its value until the LED just lights.
What is the voltage shown on the meter?

Why do we bother to turn on an LED like this when a switch will do just as well? It is because we may place a whole variety of components in the VR_1 position which change their resistance with temperature, light intensity, etc. These will make many useful automatic systems.

Task

Copy this summary into your notebook, filling in the blanks.

Transistors
Transistors have three leads called the c..., b... and ....
A small current flowing into the ... will allow a much ... current to flow through the emitter and .... To drive enough current into the base, the base-emitter voltage has to be about ... volts.

Questions

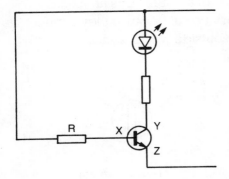

1 In the diagram the leads are

	Emitter	Base	Collector
a)	X	Y	Z
b)	Y	X	Z
c)	Z	Y	X
d)	Z	X	Y

2 In the diagram in question (1), R is there
   a) to make the base-emitter voltage 0.7 V
   b) to magnify the collector-emitter current

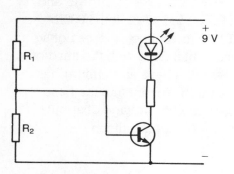

    c) to limit the base current which controls the collector-emitter current
    d) to limit the collector-emitter current which controls the base current.

3 In the diagram, the LED will light when
    a) there is 0.7 V across R_2
    b) there is 0.7 V across R_1
    c) When R_1 is twenty times greater than R_2
    d) there is 0.7 V across the LED.

4 In the diagram in question (3), if R_1 is 8400 ohms and R_2 is 600 ohms
    a) the LED will definitely light
    b) the LED will definitely not light
    c) the LED might just light
    d) the transistor might be destroyed by the larger current flowing.

Other Ways of Switching on Transistors

Transducers

A transistor uses a small base current to control a much larger emitter-collector current. The base current itself may be controlled by *light dependent resistors* which change their resistance in the dark and by *thermistors* which change their resistance with temperature. There are other devices as well. We may use these to make automatic lights, burglar alarms, heating controllers and a host of other systems. Devices like light dependent resistors (LDRs) and thermistors are called *transducers*. 'Transducer' is the name given to a device like an LDR which changes some non-electrical value into an electrical signal. Some transducers work in the reverse direction, e.g. an LED turns an electric current into light. They are very useful because they connect up electronic systems with the outside world, sensing conditions and changing things.

An LDR looks like this   This is its symbol

The resistance of an LDR may be several million ohms in the dark and less than 1000 ohms in the light.

Tasks

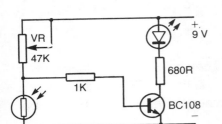

1. Switch a multimeter to the 'ohms' range. Touch the leads together. The needle should go across to the right hand side. Connect the leads to an LDR. Now cover the LDR with your hand to keep the light out and the meter reading will change.

2. Connect up this circuit. Adjust VR until the LED just goes out. Cover the LDR. What happens?

3. Rearrange the circuit like this. What can you do with this arrangement? You should find that you can adjust this circuit so that the LED comes *on* in the light.

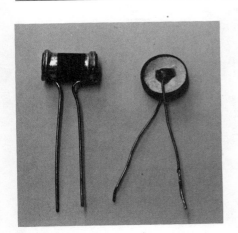

Thermistors

Thermistors are made in different sizes and shapes. Some are shown in the photograph.

Two alternative symbols for thermistors

They usually have a low resistance when hot and a high resistance when cold. (This is the opposite way round to ordinary conductors which have a higher resistance when hot.)

Task

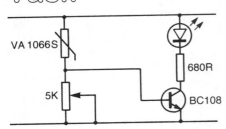

Connect up this circuit. Adjust the 5K variable resistor until the LED just goes out. You may be able to light the LED by warming the thermistor with your fingers. If not, the heat from a match will certainly make it work. (**Take care.**)

This is one way of making a fire alarm or overheating warning. The thermistor type VA1066S has a resistance of about 5000 ohms at room temperature and about 400 ohms at boiling point, 100°C. You may use other types than the VA1066S, but if you do, you may need to use a larger variable resistor than the 5K one in the diagram.

Tasks

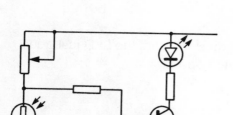

1. Try changing over the positions of the thermistor and the variable resistor in the circuit above. You may need a 47K variable resistor, but should now have a circuit which operates the LED when the thermistor gets colder.
2. Copy this summary into your notebook, filling in the blanks.

Light dependent resistors
LDRs have the symbol . . . . They have a . . . resistance in the dark and a . . . resistance in the . . . . Thermistors have a resistance which is lower when they are . . . . Their symbol is . . . .

This circuit uses the LDR to switch on the transistor and light the LED in the . . . .

Questions

1. LDRs have
   a) a low resistance in the light and their symbol is

   b) a high resistance in the dark and their symbol is

   c) a high resistance when cold and their symbol is

   d) a low resistance when hot and their symbol is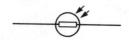

2. Which of the answers to question (1) is correct for a thermistor?

3. Which diagram correctly shows a multimeter set to read 'ohms' measuring the resistance of an LDR?
   a)   b)   c)   d)

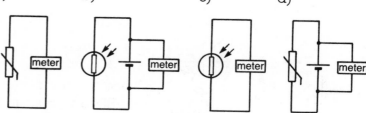

Transducers and Relays

Transducers

There are many other types of transducer used in electronics. They all connect up an electronic system to the outside world. They *either*

    take in some information or change in light, temperature, sound, etc.

Symbol for a microphone

Symbol for a record player pick up

Symbol for a motor

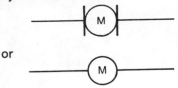

or

Symbol for a solenoid

Symbol for a transducer

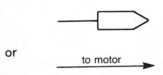

or

or

take in an electrical signal and convert it into a change in light, temperature, sound, etc.

Some transducers which change an input into an electrical signal are listed here.

Microphone   Changes sound into an electrical signal.

Record player pick-up   Changes vibrations from the record and stylus into electrical signals.

Switch   Connects or disconnects signals or circuits.

Motor   Changes electricity into movement – usually rotary movement.

Solenoid   Changes electricity into linear movement.

An iron rod is pulled into the coil by the strong magnetic effect produced by a current flowing in a coil.

Heaters   Change electricity into heat.

Sometimes you may see a transducer drawn into a circuit just as a 'black box', or as arrows leaving the circuit.

The lamps, LDRs, LEDs and thermistors we have already dealt with are transducers. What is the input and output for each of these?

Relays

Relays are really a special type of switch. They allow a cheap low power transistor to switch very high power motors, heaters, etc. They have

- a coil with a soft iron core. This becomes an electromagnet when current flows in the coil
- a moving piece of soft iron called the *armature*
- high power switch contacts which are worked by the armature.

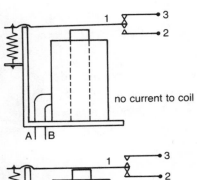

This is a diagram of a relay. The spring holds the armature so that contacts 1 and 2 are open (or not connected). When a small current is passed through the coil connections A and B, then the core becomes an electromagnet. The armature is pulled down so contacts 1 and 2 close thus making a connection.

A relay may operate several such contacts switching several things on together. It may also disconnect one circuit when it makes another. In these diagrams the relay normally connects contacts 1 and 3 all the time. Passing a current through the coil AB connects contacts 1 and 2 and disconnects 1 and 3.

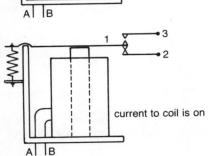

1 and 2 are called the *normally open* contacts. 1 and 3 are *normally closed* contacts.

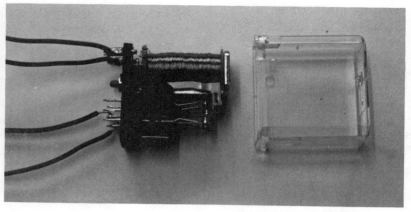

Tasks

1. Use a multimeter to find out which connections of a relay are
   a) the coil connections
   b) the normally open connections
   c) the normally closed connections.

   To do this set the meter to the 'ohms' range. Touch the leads together and the needle should go right across the dial. The coil connection should show some resistance. The normally open connections should produce no meter indication. The normally closed contacts should show very low resistance.

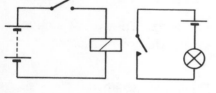

2. Use the relay to switch on a bulb by connecting it in this arrangement.

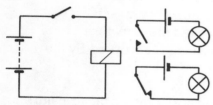

   Now connect it into this circuit so that as one bulb is turned on the other goes off.

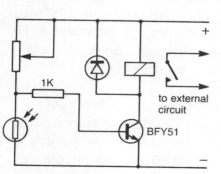

3. You can make a light-operated relay by wiring this circuit. The relay may be used to operate a motor, a lamp or any other suitable transducer. You could make a 'darkness-operated motor', a temperature-operated solenoid, etc. etc.

The diode is there to protect the transistor from a large voltage which will flow when the relay is disconnected. This voltage is called the *back EMF*.

A reed relay

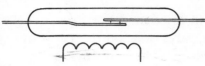

This is its symbol

Reed relays

A reed relay is a glass tube containing two metal strips. Bringing a magnet near the tube moves the metal strips so that they make contact (or break contact if they are normally closed).

Reed relays may be operated by passing current through a small coil, just like an ordinary relay.

Task

Copy this summary into your notebook, filling in the blanks.

Transducers and relays
Transducers take in information and convert it into an ... signal or they may convert an electrical ... into light, heat, sound etc.
The symbol for a microphone is ....
The symbol for a loudspeaker is ....
The symbol for a motor is ....
Relays use small currents to switch larger currents on or off.

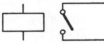 This one has a pair of ...... contacts.

Questions

From the list of answers a) to i) choose which one describes what each of the following does.
1. a bulb
2. a motor
3. a thermistor
4. an LDR
5. a loudspeaker
6. a microphone
7. a solenoid
8. a heater
9. a record player pick-up.

a) Changes electricity into linear (push or pull) motion.
b) Changes electricity into light.
c) Changes electricity into sound.
d) Changes sound into electricity.
e) Changes small vibrations into electricity.
f) Changes electricity into rotary motion.
g) Changes temperature variation into a resistance change.
h) Changes light into a resistance change.
i) Changes electricity into heat.

Fault Finding

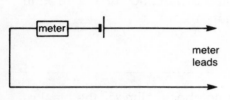

By now you will realise that when you connect up an electronic circuit things do not always go according to plan. They may not work *either* because you have not made the correct connections *or* because components are faulty. Skilful use of a multimeter solves these difficulties. The multimeter has a range marked 'ohms' or just 'Ω'. When switched to this range a battery is connected to the meter as shown in the diagram. If you hold the meter leads together the needle goes right across the dial away from the zero. This is called *full scale deflection* or *FSD*.

There is also a control marked 'ohms adjust'. Turn this to set the needle exactly on the zero at the end of the scale. Any *extra* resistance you put between the leads may now be read off the meter dial. You may use this system to measure resistance values. Digital electronic meters may zero themselves automatically. *Refer to the instructions for the particular digital meter you are using.*

Task

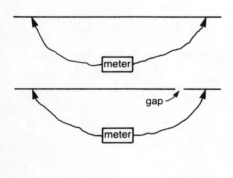

Measure five different resistors using the meter. Check that the values you measure correspond with the markings. They may not correspond exactly. Why is this?

Now think about what sort of reading your meter should give when set to read ohms when connected to a length of insulated wire. Obviously it should have a very low resistance. The meter has shown that there is *continuity* in the wire.

Suppose there was a break in the wire which was hidden by the insulation. The needle will now not move at all – it shows an *open circuit* and reveals a break in the wire. The wire has failed the *continuity* test.

Suppose we have some insulated twin flex of the kind used to connect table lamps. This should show an open circuit if the insulation is in order.

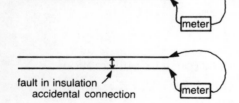

If there *is* a fault in the insulation then a very low reading will be obtained. This is a *short circuit*.

From this you can see that the meter can diagnose two faults in an ordinary piece of wire.

Here are some of the results you should get with different components.

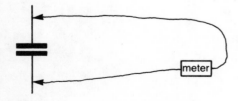

Capacitor   Open circuit – very high resistance.

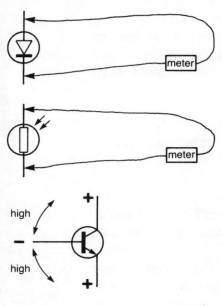

Diode High resistance in one direction. Low resistance in the other direction.

LDR Resistance which is higher in the dark than the light.

Transistor High resistance in one direction between base and collector, also base and emitter.

Low resistance in the other direction between base and collector, also base and emitter.

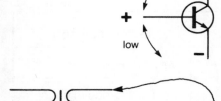

Transformer A resistance reading across the terminals of one coil. The exact value of the reading will depend on how many turns of wire there are in the coil.

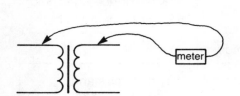

A very high resistance reading showing that there is good insulation between the coils.

If you do not get these results, the component is probably faulty. If you can get hold of some components which are known to be faulty compare them with working ones.

A multimeter set to measure ohms will also check the connection between components.

Questions

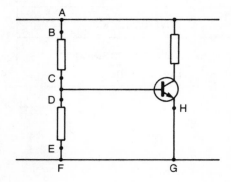

Between A and B there should be such a low resistance that you cannot measure it with an ordinary meter.

1. What resistance would you expect
   a) between C and D
   b) between E and F
   c) between F and G
   d) between E and H?

2. If connection F had not been properly made, what readings would you get
   a) between E and G
   b) between E and H?

Very often an electronic system is found not to be working simply because it is not plugged in or switched on. Just looking may be enough to check on this. But you may also use the multimeter set to read volts to check that power is reaching a circuit.

Questions

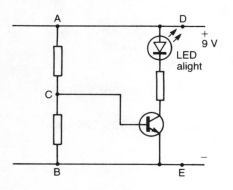

1. What voltages would you expect to find
   a) between A and B
   b) between D and E
   c) between C and B?

2. If the reading between D and E was 9 V and that between A and B was zero, where would you start looking for a fault?

3. Below are some test results using a multimeter set to read ohms. Decide from the information whether the component is *good* or *faulty*. 'Very high resistance' means greater than 1 MΩ (1 MΩ is 1 000 000 ohms).

	Component	Meter reading
a)	A 1 μF capacitor	Very high resistance
b)	A diode	Low resistance in each direction
c)	A thermistor	50 K resistance when cold, 5 K resistance when hot
d)	A transformer primary coil	30 ohms
e)	A transformer secondary coil	Very high resistance greater than 5 MΩ
f)	An 0.01 μF capacitor	Low resistance
g)	A bulb	Very high resistance
h)	A 100K resistor	998 000 ohms
i)	An LED	Low resistance in one direction, high resistance in the other direction
j)	A loudspeaker	Low resistance
k)	A length of twin flex	Very low resistance between the wires

Amplifiers

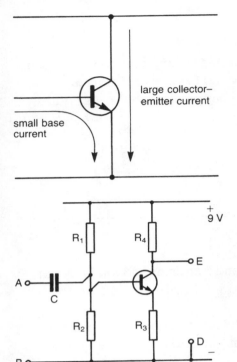

Transducers such as microphones and record player pick-ups only produce a very small electrical signal. Loudspeakers need a more powerful signal to make them work. An *amplifier* has to be connected between the two to *magnify* the small signal and *match* it to the speaker. A transistor may amplify a signal because the very small base current controls the much larger collector-emitter current. So if a changing signal is applied to the base, a much larger version of it appears in the collector circuit.

To make this happen we have to set the *bias* on the transistor exactly right. R_1 and R_2 set the base voltage. R_3 sets the voltage of the emitter. This is important in an amplifier. R_4 is the *load*. The output appears across the load.

The *input* signal from the microphone or record player is connected to A and B. The *output* is obtained from connections D and E. The capacitor C prevents any direct current reaching the transistor from the input. If DC were to reach the base it might upset the bias. But the capacitor passes only a varying (AC) signal.

Task

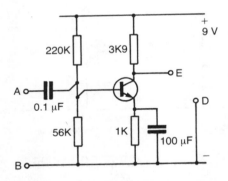

1 Connect up this circuit. Connect a microphone to A and B and a *crystal* earpiece to D and E. Any sound received by the microphone will now be amplified and sound more loudly in the earpiece. A second crystal earpiece will work as a microphone and may be connected directly across the 56K resistor.

2 Leave the amplifier connections as they are. Use an oscilloscope to measure the *input* voltage across terminals A and B when someone is using the microphone. (Just connect the Y input to A and B to do this.) Now use the oscilloscope to measure the *output* voltage across terminals E and D. You should find the output voltage is larger than the input.

Alternatively you may use a *signal generator* to make an input signal. A signal generator produces low power AC. The frequency and amplitude of this low power AC may be adjusted over a wide range by the generator controls.

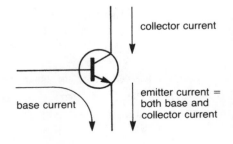

There are two other things you should know about transistor amplifiers. Firstly, the circuit we have been using is called a *common emitter* circuit. This is because both the base current *and* the collector current flow through the emitter. Other arrangements are possible, but the common emitter is the one which is used most often.

Amplifiers can be represented like this

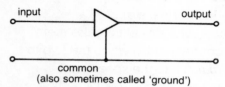

common
(also sometimes called 'ground')

Secondly, always remember that an amplifier has *two* inputs. These are the *power supply* and the *signal*. If you just connect up one then nothing happens, which is very dull!

They can also be shown like this

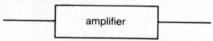

Tasks

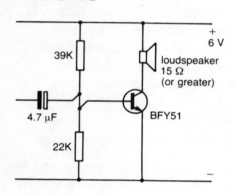

1 Try connecting up a second transistor and loudspeaker as shown.

This will give more amplification. The changing signal from the output of the first transistor drives the second one. The second transistor, a BFY51, is being used as a *power* amplifier. The final circuit now looks like this.

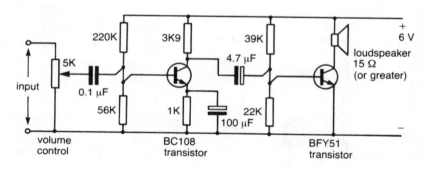

You may use another loudspeaker as the microphone in this circuit.

2 Copy this summary into your notebook, filling in the blanks.

Amplifiers
To make a weak signal drive a loudspeaker an . . . is used. The amplifier often uses transistors in the . . . emitter circuit. Amplifiers have an . . ., an output and a . . . . . . connected to them.

Questions

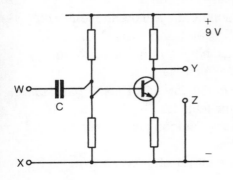

1. In this amplifier circuit
   a) a microphone would be connected to W and X
   b) an earphone would be connected to W and X
   c) capacitor C prevents base current from flowing
   d) the voltage across Y and Z must always be less than that across W and X.
2. In the circuit for question (1)
   a) an earphone would be connected to Y and Z
   b) a microphone would be connected to Y and Z
   c) no power supply is needed
   d) capacitor C prevents base current from flowing.
3. Amplifiers need
   a) a signal only
   b) a power supply only
   c) transistors which are always arranged in a common emitter circuit
   d) a signal and a power supply
4. In a transistor amplifier
   a) the output signal is a distorted version of the input
   b) the output signal is a magnified version of the input signal
   c) the amplified signal biases the transistor
   d) capacitors bias the transistor.

Integrated Circuits

You may have noticed that the transistor circuits in this book are all very similar. The common emitter circuit with its bias resistors is used in very many different systems. This means that each time one of these circuits is assembled resistors and a transistor are soldered to a circuit board. Ten or more soldered joints have to be made. Each resistor has colours on it to indicate its value and each transistor is marked with its type number.

All this soldering and all the markings could be avoided by making all the components on a single piece of silicon. The piece of silicon may have an area which is much less than a fraction of a square millimetre. A piece this size may have very many resistors, diodes and transistors etched onto it. Each is connected up automatically in the etching process. No coloured markings are needed for resistors. The transistors do not need identification printed on them. Devices made like this are called *integrated circuits* (or *silicon chips*).

Integrated circuits

An 8 pin d.i.l. IC

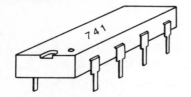

The symbol for an integrated circuit

The machines needed to design and make them are very expensive, but the final integrated circuit may be very cheap. This is because the machine may make these *ICs* by the million and all the costs of assembling individual components is avoided. ICs just have a number on them identifying the type. There are very many.

The one shown is an *8 pin dual in line IC* or just *8 pin d.i.l.* Some ICs have many more pins than this; 10, 16, 24 and 40 pin ICs are quite usual. The actual silicon part is much smaller than the packaging you see.

The symbol for an IC may be either of those shown.

We need not worry about the actual circuit *inside* an IC. In fact the manufacturer does not need to tell us much about it. All we need to know is what it will do when we apply various voltages and signals.

We now need a phrase to describe the older way of assembling circuits using separate components. We say circuits made in the old way are made from *discrete components*. (Discrete means *separate*.) Do not think that all discrete circuits are now obsolete. Look inside even the newest electronic gadget and you will see ICs and discrete components side by side. There has to be a need for many thousands of a particular circuit to make it worthwhile making an IC.

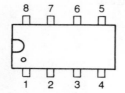

To identify the connections on an IC there is a notch cut out of one end and a dot stamped on it next to connecting pin number one. The connections are numbered as shown, as seen from above.

Tasks

1. You might have found it hard to make the amplifier in the last section. To see just how much work ICs save try making this amplifier. It only uses four components.

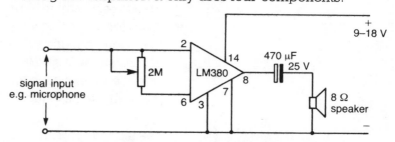

   Note The LM380 is a 14 pin d.i.l. IC. You need connect only the pins shown.

2. This integrated circuit will flash an LED on and off. You only need the LM3909 IC, an LED, a 100 μF capacitor and battery.

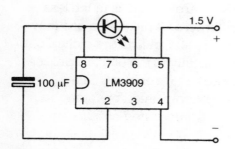

   Try using other capacitors than 100 μF. Try 1000 μF, 470 μF and 47 μF.

3 Copy this summary into your notebook, filling in the blanks.

Integrated circuits
Integrated ... may have very many components on a very small piece of .... Although the machines to make them are very ..., the final IC is usually ... because costs of making d...e components and assembling them are .... They are more reliable than circuits made from discrete ... because there are fewer soldered j....

Questions

1. ICs are made because
   a) the machines needed are very expensive
   b) the machines needed are very cheap
   c) very complex circuits may be made on a very small space
   d) it is a cheap way of making circuits when only small numbers are needed.
2. An 8 pin d.i.l. integrated circuit
   a) has eight connections and is made of discrete components
   b) has eight connections and identification letters on the outside
   c) has letters on it which tell us what components are inside
   d) has eight transistors or diodes inside it.

Logic Systems

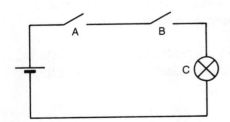

Look at the circuit diagram. The bulb will only light if switches A *AND* B are closed. A circuit of this type is called a *logic gate*. The possible combinations of switches are shown in this table.

Switch A	Switch B	Bulb C
off	off	off
off	on	off
on	off	off
on	on	on

Now let us imagine that the switches and bulb each represent either the number *nought* or *one*. Closed switches are 1, open switches 0; lit bulbs are 1, unlit bulbs are 0. The table may now be rewritten like this.

The symbol for an AND gate

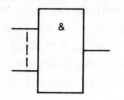

A	B	C
0	0	0
0	1	0
1	0	0
1	1	1

This is called a *truth table* for the *AND logic gate*. This type of system is used in a number of ways. For example, a machine will only issue a ticket when *A* a coin is inserted, AND *B* a button is pressed.

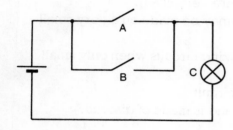

Now look at this circuit. The bulb will light if either switch A *OR* switch B is closed.

The truth table looks like this.

A	B	C
0	0	0
0	1	1
1	0	1
1	1	1

The symbol for an OR gate

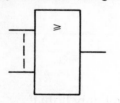

If we just used switches for these circuits they would be useful, but large and expensive. However we can now make these logic gates using transistors which are either off (logic 0) or on (logic 1). Because we can make hundreds of transistors on an integrated circuit we can fit many gates on a small IC at very low cost. This has made possible the whole range of *digital* calculators, watches and computers.

In digital systems voltages and currents rise or fall in definite steps. There are no intermediate values. It is either nothing or a maximum.

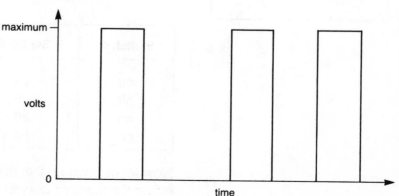

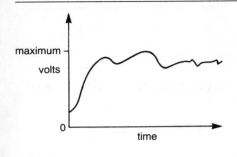

The other sort of electronic system is called an *analogue* system and the values vary continuously. Voltage may be at any value between zero and the maximum.

Another gate is an *inverter* or *NOT* gate. This just has an output which is the opposite of the input.

The truth table for an inverter is like this.

The symbol for an inverter

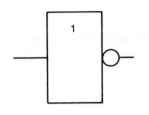

A	C
0	1
1	0

If you connect the output of an AND gate to a NOT gate the output of the whole thing will be the opposite of an AND gate.

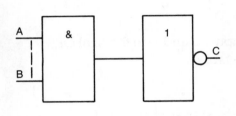

A	B	C
0	0	1
0	1	1
1	0	1
1	1	0

The symbol for a NAND gate

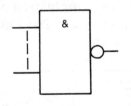

Very often in electronics AND gates and NOT gates are combined. This type of gate is called a *NAND* (NOT AND).

Likewise an OR gate with a NOT gate gives a *NOR* gate.

The truth table for a NOR gate looks like this.

The symbol for a NOR gate

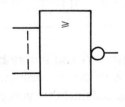

A	B	C
0	0	1
1	0	0
0	1	0
1	1	0

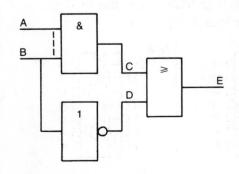

Gates are often joined together, the output from one being connected to the input of another.

To find out what E will be for different combinations of A and B we have to draw up a truth table which shows C and D as well.

A	B	C	D	E
0	0	0		
0	1	0		
1	0	0		
1	1	1		

Step 1 The first step is just to do the table for the AND gate.

A	B	C	D	E
0	0	0	1	
0	1	0	0	
1	0	0	1	
1	1	1	0	

Step 2 Now include the NOT gate. Its output will be the opposite of B.

A	B	C	D	E
0	0	0	1	1
0	1	0	0	0
1	0	0	1	1
1	1	1	0	1

Step 3 You now know the inputs C and D to the final OR gate so you can complete the table.

Try and work through the steps above for yourself.

Tasks

1 Connect up an IC type 7400. The power supply must not be greater than 5 volts or damage may result. The 7400 has four NAND gates arranged as shown.

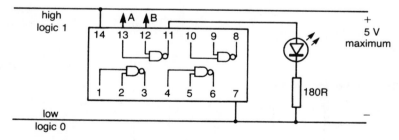

When the LED lights it indicates logic 1. A and B may be connected in various combinations with the +5 volts wire high, logic 1, or low, logic 0. You should find that you can draw up a NAND truth table from your results. You may use other ICs. The 7402 has four NOR gates; the 7404 has six inverters. The 7408 has four AND gates.

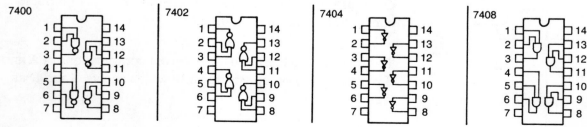

2 Copy this summary into your notebook, filling in the blanks.

Logic gates

AND gates
The symbol is   The truth table is

A	B	C
0	0	
0	1	
1	0	
1	1	

OR gates
The symbol is   The truth table is

A	B	C
0	0	
0	1	
1	0	
1	1	

NOT gates
The symbol is   The truth table is

A	B
0	
1	

Now draw the symbols and write out the truth tables for NOR and NAND gates.

3 Write out the truth table for these gates.

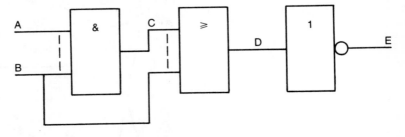

A	B	C	D	E
0	0			
0	1			
1	0			
1	1			

Questions

1. Digital signals
   a) have voltages which rise or fall only in definite steps
   b) have currents which may vary continuously
   c) are the opposite of analogue signals when passed through NOT gates
   d) can only be used in integrated circuits.
   e) More than one of the above is correct.
2. A gate which always gives a logic 1 output for two logic 0 inputs is
   a) a NOT gate
   b) a NAND gate
   c) an AND gate
   d) an OR gate.

Microcomputer Systems

Someone is using a computer. She is putting information into the keyboard and getting other information back from the screen. The computer has been given a set of instructions called a *program*. The program tells the computer what to do with the information. The program was in a *store* on a magnetic disc or cassette tape. When the program was put into the computer, the computer read what was in the store and put it into its *memory*.

This is a diagram of the system.

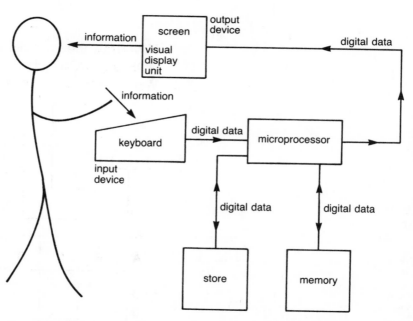

Note The person gives and receives *information*, the computer handles *digital data*.

The *microprocessor* is an integrated circuit which controls the flow of data round the circuits to and from memory, store, screen and keyboard. This microprocessor (also sometimes called a *central processing unit*, or CPU) will
- hold the data which is actually being used at that moment (this is called the *register*)
- do simple arithmetic (this is the *arithmetic logic unit* or ALU)
- control the whole system.

The store may be in a *floppy disc* as well as a tape. Floppy discs are much faster to use than tapes. They also hold much more information.

The computer handles *data* in a digital form. Each letter of the alphabet, and each number, is usually represented by a code of eight *bits*, each bit being either 0 or 1.
For example,

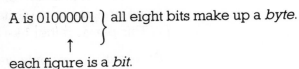

A is 01000001 } all eight bits make up a *byte*.
↑

each figure is a *bit*.

It is because cheap ICs handle this data very fast that modern computers are possible.

The data is sent between parts of the computer by connections called *buses*. These are usually wide ribbons of cable which may have 8, 16 or more wires in them. There is a *data bus*, an *address bus* and a *control bus*.

Data bus
The data bus carries data one byte at a time. So it has eight wires to keep each bit separate.

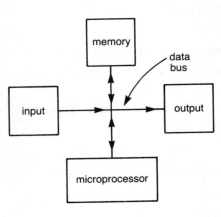

Address bus
Each part of the memory and the input and output has an *address*. This enables locations in the memory to be found again. The address is a code of 16 bits.

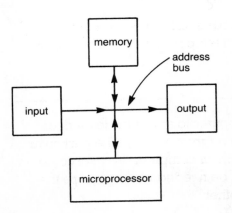

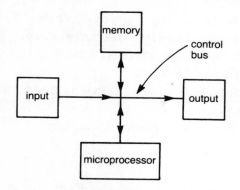

Control bus

There are several wires. Each wire corresponds to an instruction, e.g. '*read data from memory*', '*send data to output*'.

Computer memory

Imagine you have just loaded a game from a tape into a computer. It has to remember three things:

- the program itself.
- the keys which have been pressed as the game is played.
- the information needed by the microprocessor to change the program into digital signals which it understands. This part is rather like a dictionary.

Of these three, we only want the third to be remembered permanently. The first needs to be erased when we want to load another game and the second changes during the game itself as keys are pressed. So there are two types of memory.

Random access memory (RAM) This is lost or forgotten as soon as the power is turned off.

Read only memory (ROM) This is not lost when the power is turned off. The gates are permanently set by the manufacturer when the IC is made.

The ROM remembers the dictionary or interpreter the computer uses. The RAM remembers the program and what is going on at any moment.

Computers do not only have to use keyboards as inputs and screens as outputs. Other inputs may be

- a magnetic strip read from a card used in a cash machine
- a transducer which measures temperature, pressure, fuel flow, etc.

Other outputs than a screen may be

- a printer which prints words and figures for a bill
- transducers which control temperature, fuel flow, robot arms, etc.

Interfacing circuits

You cannot connect any computer to any of these devices directly. One type of printer might not accept the particular digital signals of another make of computer. To overcome this, interfacing circuits may be needed which change the signals from one form to another.

Microprocessors are increasingly used to control the fuel flow in a car engine matching its road speed and load to obtain better economy. They will also control sequences of operations in a washing machine, remember and produce complex patterns of stitches in sewing machines and operate alarms. Once set up this type of microprocessor does not need to have any change made in its program.

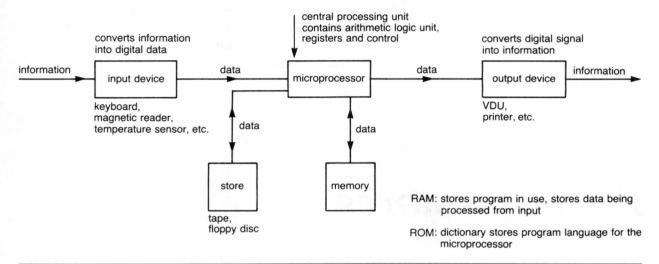

Task

Copy this summary into your notebook, filling in the blanks.

Microcomputer systems
Computer systems consist of

- an ... device which takes in i... and converts it to ....
- a memory which
  a) stores the ... in use
  b) stores data from the ...
  c) stores a dictionary of information needed by the ....
- a microprocessor which can do a ..., remember the ... in use as well as control all the rest
- an output ... which provides information, or controls things or prints
- a ... in which data is held.

The data bus carries data one ... at a time.
The ... bus carries signals which let the microprocessor find different locations in the memory or ... or output .... Each wire of the ... bus carries a signal which gives an instruction.
The ... is lost when the computer is switched off but the ... remains. The ... is a set of instructions which may be held in a .... The store may be on ... or a ...... which is much ... and will hold more than a tape.

Questions

1. When a computer is switched off, data is lost from the
   a) RAM b) ROM c) magnetic tapes d) floppy discs.
2. The operator can never change the data in a
   a) RAM b) ROM c) store d) bus.

3 The program is a set of instructions which
   a) tells only the memory what to do
   b) is held on a magnetic tape or a floppy disc
   c) is entered only through a keyboard
   d) tells only the output device what to do.
4 The computer handles data in digital form
   a) because this is the only way magnetic tapes work
   b) produced by the central processing unit
   c) only when it is in the data bus
   d) using a code where combinationss of 0 and 1 represent letters and numbers.

Audio Systems

Amplifiers

An *audio system* produces sound from a signal. To do this it uses an amplifier to drive a loudspeaker.

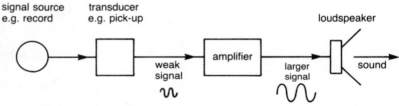

The *signal source* may come from a radio, a cassette tape *head*, a record *pick-up* or a microphone.
All these sources only produce weak signals. They need amplifying before they will make a loudspeaker work.
  The amplifier in an audio system has two main sections. These are the *preamplifier* and the *power amplifier*.

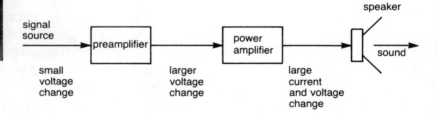

The *preamplifier* amplifies the very small voltage change from the source and makes it into a larger voltage change. The *power amplifier* changes the signal from the preamplifier into a still larger voltage change with a much larger current flowing.

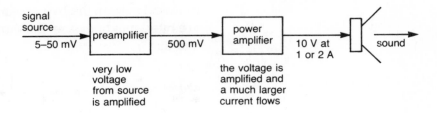

The preamplifier does more than just amplify. There are *controls* on it which let us adjust the sound.

Volume This adjusts the loudness of the sound by changing the amplitude of the signal.

Bass This adjusts the amplification of the low pitched sounds (low frequencies).

Treble This adjusts the amplification of the high pitched sounds (high frequencies).
The bass and treble are the **tone controls**.

Loudspeakers

The *loudspeaker* is a large cone of material which is pushed rapidly in and out to make the sound. The cone is pushed to and fro by the vibrations of a small coil set inside a powerful magnet. The coil vibrates because it is connected to the varying signal current from the amplifier. This current makes a magnetic field round the coil. Since the current changes, the magnetism changes and the coil moves.

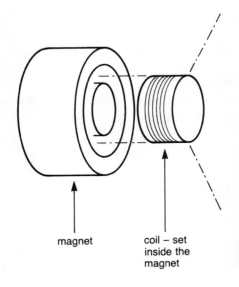

Some people can hear a range of sounds with frequencies between 20 Hz and 20 000 Hz, but most people hear sounds between about 40 Hz and 15 000 Hz. The actual figures are different for different people. There are no hard and fast rules. The speaker cone has to vibrate at these sound or *audio* frequencies. So do the electrical signals from the source right the way through the amplifier.

Stereo systems

Most audio systems are *stereophonic* or simply *stereo*. This gives a more satisfying sound to music and drama. A stereo system needs two loudspeakers, two power amplifiers, two preamplifiers and two separate signals. Of course, in a practical system the two preamplifiers and power amplifiers are fitted into one case. But our system diagram now looks like this.

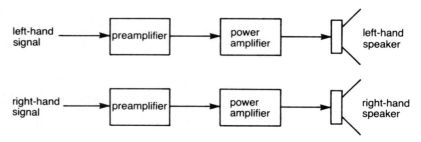

Task

Copy this summary into your notebook, filling in the blanks.

Audio systems
Audio ... use a transducer like a m... or a record p...p to produce a signal. This signal is amplified in a ... amplifier which also has ... controls and a volume .... The signal from the pre... drives the ... amplifier which produces a large power change. This operates the loudspeaker. Amplifiers need a power s...y as well as a signal in order to work.

Questions

1. The sequence through which a signal passes in an audio system is
   a) power amplifier, pick-up, preamplifier, loudspeaker
   b) transducer, preamplifier, loudspeaker, power supply
   c) transducer, power amplifier, preamplifier, tone control
   d) transducer, preamplifier, power amplifier, loudspeaker.
2. Which one of the following is *not* connected to the power supply?
   a) power amplifier
   b) loudspeaker
   c) right hand preamplifier
   d) left hand preamplifier
3. A preamplifier
   a) increases the output current
   b) increases the input voltage to the power amplifier
   c) may have tone controls
   d) is only used for stereo amplifiers.
   e) More than one of the above is correct.
4. Stereo amplifiers
   a) need two speakers, but only one preamplifier
   b) produce sounds between 40 and 15 000 kHz
   c) need two speakers, two amplifiers and preamplifiers and two input signals
   d) have two amplifiers both running from the same identical signal.

Radio Receiver Systems

Radio waves and sound waves

This British radio receives signals from all over the world at frequencies up to 30 mHz

You might be able to hear a record player playing at normal volume 50 metres away. If you turned it up really loudly, you might treble this distance. Yet with a radio you may hear signals ten thousand times further away than this. What is it that gives a radio this enormous range? It is because the *radio waves* are totally different from the sound waves we hear from a speaker.

Sound waves travel through the air. They have a speed of about 340 metres per second (or 1200 kilometres per hour, or 750 miles per hour). The frequencies we can hear have an upper limit of about 15 000 hertz for most people. (15000 Hz may be written 15 kHz which is 15 kilohertz.)

Radio waves may have frequency as high as 100 000 000 000 hertz (or 100 000 megahertz, 100 000 MHz for short). Even if your ear could hear this sort of frequency you would not be able to hear radio waves. This is because they do not *themselves* produce sound waves. Radio waves are *electromagnetic waves*. They have enormous speed of nearly 300 000 000 metres *per second*. This is the speed of light. (How far do they travel in an hour?)

A radio wave could travel seven times round the earth in 1 second. A sound wave would take over 33 hours to go round once.

Receivers

A radio set has to *detect* these radio waves and extract the sounds put onto the waves at the transmitter. It also has to *tune* in the signals, separating out the one you want to listen to from the many thousands being transmitted.

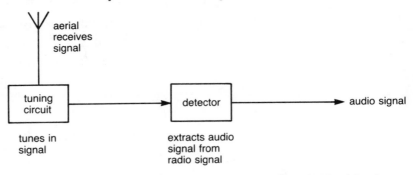

To complete the very simple receiver system in the block diagram we have to add an *amplifier* as the signal is very weak, also a *transducer* to turn the signal into sound.

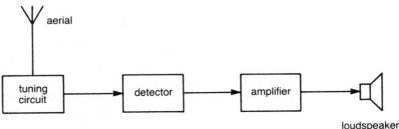

Task

Connect up this circuit.

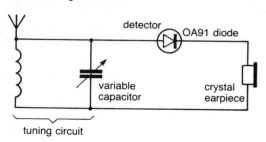

The symbol for a variable capacitor

The coil should have about 70 to 80 turns on a cardboard tube about 4 cm in diameter.

You should find that you can receive a few local stations on this *crystal set*. You *must* have a long aerial to make this work – 20 metres or more – high up. Also you *must* use a crystal earpiece or 'high impedance' earphones. Ordinary stereo headphones will not work at all in this circuit.
The photograph shows a variable capacitor. Old radio sets are a good source of these components.

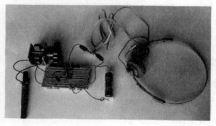

This photograph shows an *integrated circuit receiver*. Note the ferrite rod on the left; also the way in which the headphones are connected.

Tasks

1 Make up this receiver using an integrated circuit type ZN415E or 2N416 which gives a louder signal. It will give a good performance receiving signals between about 500 kHz and 1500 kHz. The coil should have about 80 turns wound onto a piece of ferrite rod.

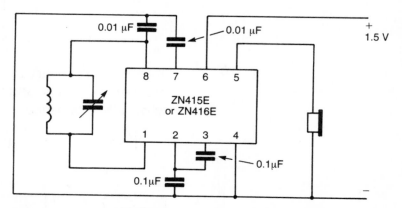

The variable capacitor must be close to the IC. Use stereo headphones with only the tip of the plug and its ring connected.

2 Experiment by winding a coil with about 60 turns. You will find you receive stations on *higher* frequencies or shorter wavelengths. If you add turns to the coil you will receive stations on *lower* frequencies or longer wavelengths. How many turns will just receive BBC Radio 4 on long wave?

Frequency and wavelength

Radio dials are marked in frequency or wavelength. To understand the connection imagine a swimmer standing in the sea. Waves come in at 1 metre per second. They are 1 metre from crest to crest. This is their *wavelength*.

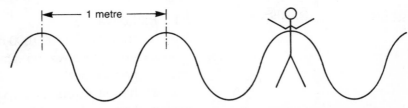

How many pass the swimmer each second? The answer is 1 per second. This is their *frequency*.

Now look at another swimmer standing in the waves. Their speed is still the same, but now they are half a metre from crest to crest. Their wavelength is 0.5 m.

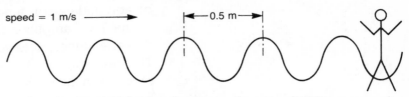

How many pass this swimmer each second? The answer is two per second. Their wavelength is less, but the frequency is greater.

Question

What would be the frequency if the wavelength was 2 metres? What would be the wavelength if the frequency was 4 Hz?

Water waves are even slower than sound, but the connection is the same for all of them.
  speed = frequency × wavelength.

Since the speed of radio waves is constant, if the frequency gets larger the wavelength gets smaller, and vice versa.

AM and FM

Modulation is the process of making the radio wave carry a signal. With *amplitude modulation* (*AM*), the amplitude of the wave is varied. With *frequency modulation* (*FM*) the frequency is changed. AM is used on long and medium waves. FM is used on **v**ery **h**igh **f**requency or *VHF*.

Question

Would you expect long waves to have high, medium or low frequencies? What sort of wavelength do VHF signals have?

Task

Copy this summary into your notebook, filling in any blanks.

Radio receivers

Radio sets need an ... to pick up the electromagnetic wave. The signal is sorted out from the number being transmitted by the .... The ... removes the audio ... from the radio signal. The ... magnifies this signal and the ... turns it into ....

Questions

1. Radio waves
   a) travel at 300 000 000 metres per second
   b) are sound waves which travel at 300 000 000 metres per second
   c) have frequencies up to 15 kHz
   d) have frequencies up to 100 000 MHz.
   e) Both a) and c) are correct.
   f) More than one of the above are correct.
2. The order in which a radio signal passes through a receiver is
   a) aerial, tuner, amplifier, detector, speaker
   b) aerial, detector, tuner, amplifier, speaker
   c) speaker, amplifier, detector, tuner, aerial
   d) aerial, tuner, detector, amplifier, speaker.
3. The detector
   a) tunes in the signal
   b) converts the radio signal into sound
   c) converts the radio signal into an audio signal
   d) amplifies the radio signal.
4. The tuner circuit is necessary because
   a) the audio signal has to be extracted from the radio signal
   b) radio signals may be very weak
   c) the detector will not work without one
   d) there are many thousands of signals being transmitted.

Feedback Control Systems

Every day of our lives we get *feedback* about things. Is it too cold? Then we get up and turn the heating on. If it gets too hot, we turn it off again. When cycling or driving we turn to avoid things. Have we turned too far? Or not enough? If we have then we detect the error and correct it very quickly.

We even get some feedback from people's faces when we are talking to them. It shows us if they are bored, interested, amused, angry, etc.

In engineering feedback is used all the time to control and regulate.

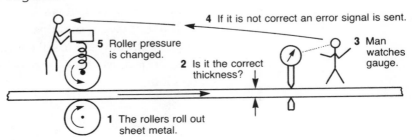

In the diagram, the *error signal* may be produced by a man watching a gauge. He then adjusts the pressure on the rollers. Or, much more likely these days, it is an automatic electronic system which constantly measures and makes adjustments in a fraction of a second.

All these feedback systems have a *feedback path* along which an error signal flows to make adjustments to something.

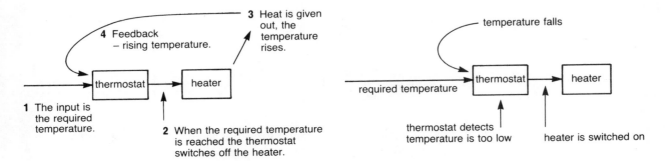

Another feedback system is a thermostat which controls a heater. This uses the temperature of the air as a *feedback path*. The *input* we provide is the setting of the thermostat to the temperature we need. The difference between the temperature we set and the actual air temperature provides an *error signal*. It may not be very quick acting and the temperature may fall to a much lower level before it switches on again. We can make this faster acting by using a thermistor to detect temperature changes and an amplifier like those on pages 62 and 64.

In practice more complicated amplifiers would be used. They are called *error amplifiers*. They magnify very slight changes so making the system very sensitive and quick acting. Our system now looks like this.

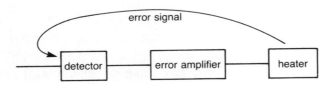

81

Another feedback control system governs the speed of electric motors. In record players and cassette tape players the speed of the record or tape has to be exact and must not change. If it is too fast the pitch of the sound from the speaker is too high and vice versa. If the speed varies, then the sound has an unpleasant warbling quality.

In old steam engines and clockwork gramophones *governors* were feedback control systems. If the motor speeded up, the weights would fly out and reduce the speed again.

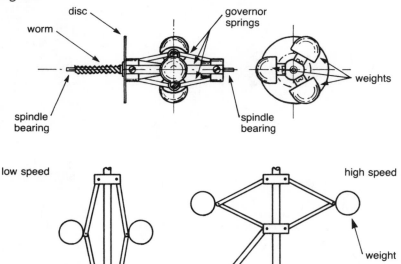

In a cassette player the tape has to move at exactly 47.5 millimetres each second (47.5 mm s^{-1}) The motor which drives the tape along does this by having its speed controlled by feedback. It works like this.
1 The motor shaft turns.
2 Its speed is measured by a circuit, or a transducer.
3 The signal from (2) is compared with a reference signal.
4 If the signal from (2) is different from the reference, then this error signal is amplified.
5 The amplified error signal is used to change the motor speed.

If the motor *slows down* then it has to be *speeded up* to correct things. If the motor *speeds up* then it has to be *slowed down* to correct things.

Note If the change is in one direction then to correct things a change has to be made in the opposite direction. This is called *negative feedback*. Negative feedback always makes for greater *stability*. It is used a lot in electronics. The opposite is *positive feedback* which makes for *instability*. Here any change in the output is amplified and produces more change, which is amplified again, and so on. A

common example of this is the howling noise you sometimes hear when the microphone and speakers of a public address system are too close.

Task

Copy this summary into your notebook, filling in any blanks.

Feedback

Feedback ... are used to control and regulate. In a motor speed controller, the speed is measured by a .... If this speed is different from the one required, then an ... signal is produced. This error signal is ... and is used to correct the motor speed.

Questions

1 Feedback systems
   a) use an error amplifier to make motors turn very fast
   b) always make electronic devices less stable
   c) are used to make amplifiers show up errors
   d) use error amplifiers to magnify a signal which controls a system.

2 A feedback control system is used to maintain a motor speed at exactly 600 revolutions per minute (r.p.m.). If the motor slows down to 580 r.p.m.
   a) an error amplifier is produced which speeds up the motor
   b) an error signal is produced, amplified and it speeds up the motor
   c) an error signal is produced, amplified and it slows down the motor
   d) the feedback path is controlled by the error amplifier to speed up the motor.

3 In a feedback system
   a) the error signal is produced by the transducer and amplified by the output
   b) the output comes from the feedback path
   c) the error signal goes through the amplifier to the feedback path
   d) the feedback path provides information about the output to the error amplifier.

Sources of Danger

Mains electricity

The mains electricity supply is dangerous. If current passes through you it may have serious, even *fatal* results. To understand how to avoid the risk of shock we have to know a little more about how electricity is generated. As we saw on pages 33 and 45, alternating current is easily made by moving a magnet near a coil. Power stations use this same principle, except that the generator coils are huge. With AC (alternating current) we cannot label a wire + or − because it is continually changing.

```
        +              −              +
        −              +              −
1              2  0.01 s  later   3  then        4  and so on . . .
```

With mains AC, *one* of these wires is connected to the earth at the generator end. So our two wires above could now be drawn like this.

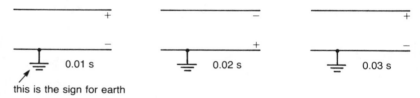

this is the sign for earth

Even though we still cannot say which is + or − as they are continually changing, we may label them *live* and *neutral*.

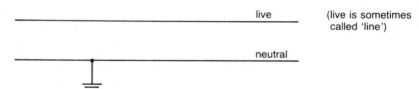

(live is sometimes called 'line')

This means that you may get a shock in two ways
- by touching live and neutral wires
- by touching a live connection and having a good connection to earth yourself.

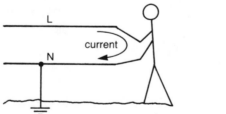
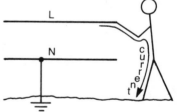

Safety devices

A serious electric shock happens when current passes through your body. As little as 10 mA (0.01 A) may be fatal. The size of the current depends upon two things. These two things are
- the voltage
- your resistance.

In fact this is just like any other electrical circuit which obeys Ohm's law. If your skin makes a good connection with the wire or contact, then the resistance is low and a larger current flows. This happens especially when your skin is moist or wet. If you are making a good connection to the earth by standing on damp soil in bare feet, for example, you are especially vulnerable. For these reasons, it is important to use safety devices.

Task

Set a multimeter to read 'ohms'. Take hold of the metal lead ends in each hand. What is the resistance between your two hands? Compare your reading with other people's. Moisten your hands with water and repeat the experiment. You will probably find the resistance is anything between 2000 and 10 000 ohms. It is low if your skin is moist and high if it is dry.

How may we eliminate the chance of an electric shock? There are safety devices built into the system which are completely effective.

The earth wire

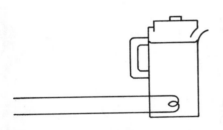

This diagram shows an electric kettle working normally. The live and neutral wires are connected to the heater element. This is kept clear of the metal body of the kettle by insulation.

The photograph shows the element heating wire inside its containing tube which has been cut away. If for any reason the live wire was to make contact with the metal body, then the kettle itself would become 'live'.

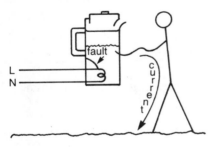

Someone touching the kettle would now act as a path for this *fault* current to flow down to earth and would get a severe electric shock.

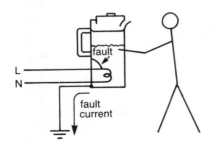

Now if we add a *third* wire connecting the kettle *body* to the earth, we make a separate route for this fault current to flow to earth. It will not now flow through the person who touches it. This is because it has a much lower resistance path through the earth wire.

To use the earth wire system properly we have to have a colour code for mains cables and to know where to connect these cables on a plug.

The code is

- the *live* wire is *brown*
- the *neutral* wire is *blue*
- the *earth* wire is *green with a yellow stripe*.

There are several ways of remembering which wire is which. One is to note neutral: blue – the only ones with '**u**' in them. The earth has green grass growing on it (and yellow flowers). So the live brown wire is the one which remains.

If you can find a memory aid which suits you better, then please use that one.

Wiring a plug

The wires are connected to the plug like this.

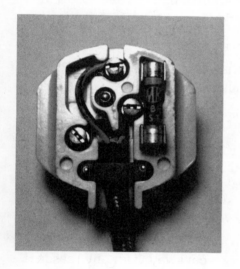
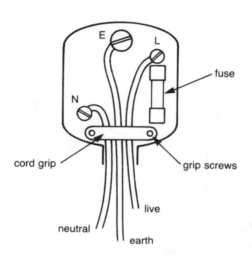

The *gripscrews* (also sometimes called the *cord grip*) are there to prevent the connecting wires being pulled out of their connections. The *whole* of the outside of the cable should be clamped under this, *not* just the three connecting wires.

Task

Copy this summary into your notebook, filling in any blanks.

Electric wires

The two wires which carry current in a mains supply are ... and .... In an electrical device working normally the live and neutral ... do ... touch the body of the device. They are i...d from it. In the unlikely event of the live wire accidentally touching the body of the device, the 'fault' current will flow down to ... through the ....

The colour code for cables is

- live ...
- neutral ...
- earth ...

Questions

1. The wires of an AC supply
   a) are live and earth
   b) are + and neutral
   c) are neutral and earth
   d) are live and neutral.

2. You are most likely to get a serious electric shock
   a) if you touch a live wire when you are in the garden in the rain
   b) if you touch a neutral wire when you are in the garden in the rain
   c) if you touch the live wire with dry hands standing on a rubber mat
   d) if you touch the earth wire.

3. The earth wire of a plug connected to an electric iron
   a) takes current to the iron from earth
   b) might sometimes take current from the iron to earth
   c) always takes current from the iron to earth
   d) never takes current from the iron at any time to earth.

4.

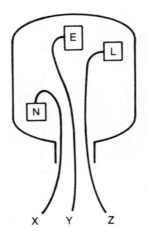

The correct wiring for the plug is

	X	Y	Z
a)	blue	brown	green and yellow
b)	green and yellow	brown	blue
c)	brown	blue	green and yellow
d)	brown	green and yellow	blue
e)	none of these.		

We shall look at some more safety devices in the next chapter. Work at the questions above until you can answer them all correctly. You must not be satisfied with anything less where safety is concerned.

Fuses and Residual Current Devices

Safety devices

Fuses and residual current devices (or RCDs) are two more very important safety devices.

Fuses

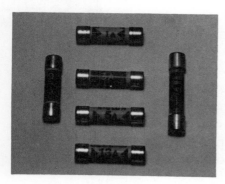

This is a fuse

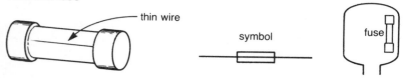

13-amp plugs have fuses in them. Fuses are weak links which have been deliberately put into the circuit. They are pieces of wire which may melt and break the circuit if the current is over a given value. You choose which fuse to put into a plug. Common fuse values are 2 amps, 3 amps, 5 amps, 10 amps and 13 amps. In a house there may also be fuses, where the cable enters the house, which melt at 30 amps.

They work like this. The fuse is in the *live* lead to an appliance. If a fault occurs, then the fault current should melt the fuse and break the circuit. The sequence, which happens very fast, is

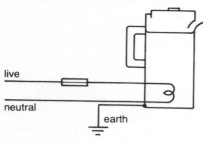

- the fault occurs and the live wire is connected to the kettle body
- a large current flows to earth
- the fuse blows and disconnects the current.

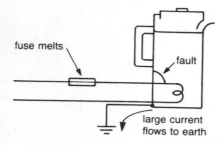

How do you know which size fuse to use? You use the formula for *power*, on pages 41 and 42.

watts = volts × amps.

If the kettle is labelled 2000 W, 240 V, we can calculate the current it takes. We use a fuse which will carry *at least* that current. In this case

watts   = volts × amps
2000   = 240 × amps
$\frac{2000}{240}$ = 8.3 amps

So you would use a 10 amp fuse.

Question

Work out which fuses you would fit to a plug for each of the following. *Remember* you may only choose 2 A, 3 A, 5 A, 10 A or 13 A fuses
a) a 750 W 240 V motor
b) a television set labelled 60 W, 240 V
c) a 3000 W, 240 V water heater
d) a 2400 W, 240 V heater

Note The answer to the last question was 10 A, but a 10 A fuse will not do. This is because 10 amps is the current which *just* makes the fuse blow. So you have to use a 13 A fuse.

The RCD

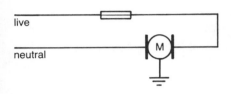

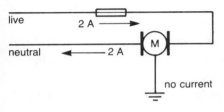

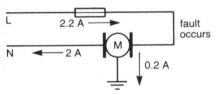

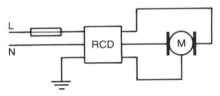

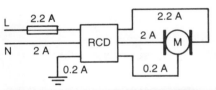

Residual current devices or RCDs are very effective safety devices. The diagram shows the motor of an electric drill working normally on 2 amps.

What can you say about the current passing along live and neutral wires? Should there be any current flowing in the earth wire? If you think about it for a while you should quickly come to the conclusion that there is *no* current flowing in the earth wire. It is not quite so easy to realise that the current in live and neutral wires must be equal. (Remember from the section on 'What is Electricity' that one wire *pushes electrons out* round the circuit and the other *receives* them.)

If a fault develops then *some* current will flow down the earth wire so the current in the live and neutral is no longer equal. The RCD detects the difference and *immediately* switches off the current. It works very much faster than a fuse.

Can you see a second reason why an RCD provides a greater degree of safety compared with a fuse? The answer is that a fault which allows only 0.2 A or less to flow through the earth wire will turn off the current. But a fuse for this motor would need at least 3 A to flow before it cut off the current.

The RCD is reset by operating the switch after the fault has been mended. The photographs show a residual current device which you may fit to a socket, and a permanently installed one.

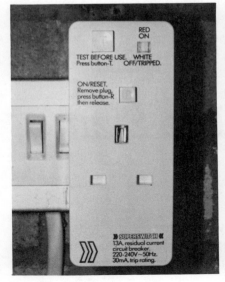

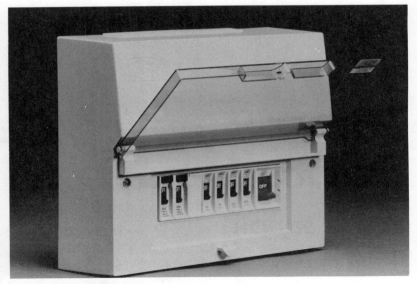

Residual current device

Task

Copy this summary into your notebook, filling in any blanks.

Fuses and RCDs

A fuse melts if the ... flowing is over a given value. The fuse is always in the ... lead of the appliance. Fuses are made in different current ratings. You have to know the power of the device in w ... and the ... it is to run on to calculate fuse size. You use the formula ... = ...

A r ... current ... cuts off the supply if the current flowing in the ... and neutral wires is not ... . It works very quickly and may be operated by a very ... current.

Questions

1. A fuse
   a) is a thin piece of wire put in the neutral lead
   b) is a thin piece of wire which melts if the current is less than a certain value
   c) melts if the current in the live wire is above a certain value
   d) switches off the supply if the current in live and neutral leads is not equal.
2. A residual current device
   a) switches off the supply if the current in live and neutral leads is unequal
   b) operates very quickly on very low currents
   c) unlike a fuse, does not depend on the total power in the circuit.
   d) All of the above are correct.
3. A motor is labelled 240 V, 500 W. The most suitable fuse would be
   a) 2 A b) 3 A c) 5 A d) 10 A
4. Fuses are always put in the live lead because
   a) there is a large potential difference between live and earth
   b) there is a large potential difference between neutral and earth
   c) most current flows in the live lead
   d) the current in the earth wire is very small.

Power Supplies, Cable Sizing, Device Handling

Power supply units

Most electronic circuits and systems run on low voltages (less than 20 V) even though they are plugged into the mains (240 V). The *power supply unit* steps down the voltage from 240 V to what is required. It also rectifies the AC and smoothes it, as described on pages 36 and 38.

Cable sizing

This is a commerical PSU

Inside a student's PSU

In the photograph of the inside of a power supply unit (PSU) you may see several safety features. The mains cable going in is fixed by a clamp. This is so that it cannot be pulled away from the transformer inside. Nor can the earth wire become disconnected from the steel cabinet. You can also see that the wires inside are of different thicknesses. This is because different wires may carry very different currents. If we pass a large current through a thin wire it may become hot – in fact hot enough to start a fire. It is very important to use a wire which is thick enough to carry the current which will flow in a circuit.

Question

The diagram shows, twice full size, some cables sold for wiring.
They are the correct sizes for currents of
a) 6 A   b) 1.4 A   c) 0.5 A   d) 20 A   e) 0.4 A   f) 3 A

Decide which wire you would use for each current.

The reason why we do not use the thickest cable for everything is because of *size* and *cost*. The 20 A cable costs ten times as much as the cheapest wire.

Never try to guess the current carrying ability of a cable from its total thickness including the insulation. Some cables have very thick insulation to withstand high voltages. Some have thin insulation because they are only intended to carry low voltage. One of these cables will carry current at 25 000 volts. Another will carry 1 amp at no more than 60 volts. Which do you think is which?

Device handling

When testing anything electrical which runs from a high voltage *always* disconnect it first. Even when switched off and disconnected, capacitors may stay charged up and give a shock. The one shown in the photograph above has a resistor across it which discharges it when disconnected. But don't assume they will always have this.

On a printed circuit board, insulation depends partly on the distance between the tracks. If the tracks are close, then they will not be able to withstand high voltages. As well as the user, components may be damaged by wrong handling. Meters, microphones, speakers and radios are easily damaged by dropping. Transistors, ICs and diodes may be damaged by overheating when soldering them into a circuit if the soldering iron is left in contact for too long. A hot iron applied quickly is best. You may protect the component by holding its lead with a pair of long-nosed pliers or a crocodile clip while it is being soldered. Then the heat is conducted along this *heat shunt* and not into the component. When soldering, also make sure the iron body does not touch nearby components. They may melt.

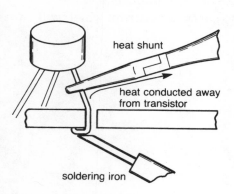

On page 29 we saw how static charges are produced by friction. Some sorts of transistors and ICs are damaged by static electricity. These should be handled as little as possible. A soldering iron with a *low leakage current* is usually used to solder these into a circuit.

Failed components

Almost all failed components may be thrown away in a dust bin. (Transformers may have a scrap value because of the copper wire in them.) Ones which need special care are

- a few types of transistor which are designed to handle large powers at high frequency. They must *never* be cut open because of the substances they contain.
- cathode ray tubes, which have a vacuum in them. They must not be removed from their casing because if struck they might implode violently.

When printed circuit boards are being made either ferric chloride or an acid is usually used for etching the copper. Both are very damaging to clothes, skin and eyes. Proper eye protection should always be worn when handling these. For disposal of exhausted chemicals, local regulations about pouring things into the drains must be observed. The local water authority will advise on this. County authorities may be contacted for information on disposal of components or substances about which there is any doubt.

Task

Copy this summary into your notebook, filling in any blanks.

Device handling
Mains cable are secured by a ... where they enter a container. This is to prevent the wires being ... away from the components inside. Thicker ... carry larger ....
A large ... in a ... wire may cause the wire to get very ....
... may remain charged even when disconnected. Care must be taken not to over ... components when soldering.

Questions

1. Cable clamps prevent
   a) wires pulling out
   b) short circuits because wires have been pulled away
   c) open circuits.
   d) More than one of the above is correct.
2. Thicker cables are used for large currents because
   a) they get warmer
   b) they are no more expensive than thin wires
   c) the cable clamp holds them in more easily
   d) thin cables might overheat.
3. Even when disconnected, care should be taken with electronic apparatus because
   a) of chemicals which were used to make parts of it
   b) transformers are always very heavy
   c) capacitors may remain charged
   d) capacitors are always discharged.
4. When soldering a transistor, the leads are sometimes held by a heat shunt because
   a) it protects the transistor from becoming too hot
   b) it makes the lead hotter so melting the solder more quickly
   c) it protects the transistor from static electricity
   d) the heat shunt has to get hot for proper soldering.

To Sum Up

Formulae and symbols

The formulae you need to answer questions on basic electronics are

watts = volts × amps

$$\frac{\text{watts}}{\text{volts}} = \text{amps}$$

$$\frac{\text{watts}}{\text{amps}} = \text{volts}$$

The triangle of volts, amps and ohms is

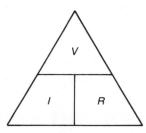

$$I = \frac{V}{R}$$

$$R = \frac{V}{I}$$

$$V = IR$$

The symbols for components we have used are

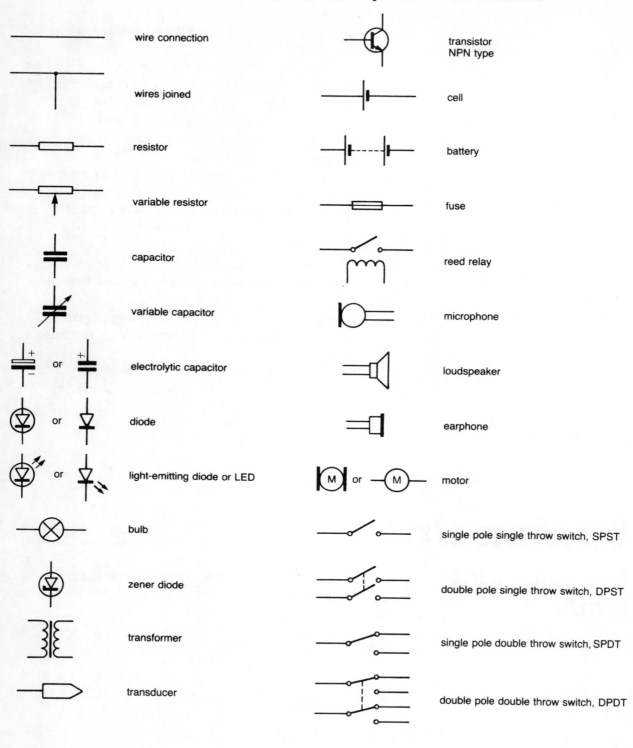

Please note that although we have only used NPN transistors there are other types.

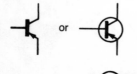 This is a PNP transistor.

This is a *Field Effect Transistor* (FET).

Components Required

The following components are needed for each student (or pair of students) to do the practical work in this book.

- light-emitting diodes (2)
- silicon diodes (4)
- zener diodes: a selection – 4V7, 6V2, 6V8, etc
- transistors BC108, BFY51
- diode type OA91
- integrated circuits 7400, 7402, 7404, 7408, LM380, LM3909, ZN415E or ZN416E
- thermistor VA1066S
- light dependent resistor ORP12
- capacitors 1000 μF, 470 μF, 100 μF, 47 μF, 4.7 μF, 0.1 μF, (2), 0.01 μF (2)
- resistors 100R, 120R, 150R, 180R, 220R, 270R, 330R (2), 680R, 1K (2), 2K2, 3K9, 22K, 39K, 220K
- variable resistors 5K, 47K, 2M
- loudspeakers 15 ohm, 8 ohm
- crystal earpiece, crystal microphone, stereo headphones
- variable capacitor and ferrite rod
- multimeter
- 2.5 V bulb in holder
- holder for up to six 1.5 V cells
- Verobloc or similar plug-in circuit board
- magnet
- about 20 m thin insulated wire
- soft iron cores, Unilab type preferred (2)
- short leads with crocodile clips and/or 4mm plugs on the ends (4)
- relay, any low current type; those with change-over contacts are most useful
- low voltage AC power supply
- signal generator (optional)
- oscilloscope, Unilab type preferred (one could be shared)

Note If a low voltage power supply is used instead of cells, then the power supply must have a smoothed output.

What next?

The Basic Skills Electronics examination gives you a broad introduction to most important areas of the subject. You may continue with it for a job or as a hobby. If you do neither of these you are still very likely to use some of the things you have learned by doing this course.

If you want to continue it as a hobby then read a monthly magazine such as *Everyday Electronics*. You may obtain components from companies who advertise in these magazines.

Rapid Electronics Ltd specialise in providing components to schools and colleges.

Alternatively some companies such as Cirkit Distribution Ltd and Maplin Electronics publish large catalogues which may be obtained from some newsagents. You will learn a lot more just by looking through the catalogues and such companies give very quick service sending components by post.

Useful addresses
Cirkit Distribution Ltd
Park Lane, Broxbourne, Hertfordshire, EN10 7NQ;
Everyday Electronics
6 Church Street, Wimborne, Dorset, BN21 1JH;
Maplin Electronics
PO Box 3, Rayleigh, Essex, SS6 2BR;
Rapid Electronics Ltd
Hill Farm Industrial Estate,
Boxted, Colchester, Essex, CO4 5RD.

Answers

Circuits
Page 6 1 b) 2 a) 3 d) 4 c) 5 a) 6 b) 7 c)

Light-emitting Diodes
Page 8 1 b) 2 a) 3 c) 4 c) 5 d)

Practical Ways of Joining Components and Circuits
Page 12 1 d) 2 a) and b) 3 b)

Series and Parallel Circuits
Page 14 1 c) 2 b)

Using Ammeters
Page 16 1 d) 2 b) 3 c) 4 d) 5 a) 6 b)

Measuring Voltage
Page 19 1 c) 2 c) 3 a) 4 d)

Resistance
Page 21 a) 4 A b) 1 A c) 2 A d) 1 A
Page 23 1 1.5 A 2 10 V 3 50 ohms 4 9 V

Resistors
Page 27 1 a) 3900 Ω b) 39 000 Ω c) 47 Ω
  d) 2 200 000 Ω e) 8200 Ω
2 a) 2K7 b) 1M c) 1M2 d) 5K6
  e) 18R f) 33K
3 a) brown green orange
  b) brown green red
  c) brown green brown
  d) brown black green
  e) orange orange yellow
  f) brown red black
4 a) 22 000, 10% b) 560, 10% c) 12, 5%
  d) 33 000, 5% e) 4700, 10% f) 27 000, 5%
5 d) 6 d) 7 c) 8 a) 9 a) 10 c) 11 b)
12 d) 13 d)

What is Electricity?
Page 30 1 b) 2 c) 3 a) 4 d)

The Cathode Ray Oscilloscope
Page 33 a) 6 V b) 5 ms c) 12 V
Page 34 1 a) G b) F c) E d) C
  e) D f) A g) B h) I i) H
2 a) 3 b) 4 d)

Rectifying AC to Produce DC
Page 36 1 c) 2 a) 3 c) 4 c) 5 a)

Capacitors
Page 40 1 a) 2 c) 3 a) 4 a)

Electrical Power
Page 42 (top) 1 120 W 2 36 W 3 375 W 4 1500 W or 1.5 kW 5 1920 W or 1.92 kW 6 0.003 W or 3 mW
Page 42 (middle) 1 1 A 2 0.5 A
Page 42 (bottom) 1 a) 72 W b) 0.5 W c) 0.54 W
  d) 0.25 A e) 0.4 A
2 2000 mW; 200 mW; 20 mW
3 2.5 kW; 0.25 kW; 5 kW
4 1.5 W; 0.5 W; 1200 W; 500 W

Transformers
Page 45 1 c) 2 b) 3 a)

More about Diodes
Page 48 40 ohms
Page 48 (bottom) 1 c) 2 b) 3 d) 4 b) 5 a) 6 c)

Transistors
Page 51 1 d) 2 c) 3 a) 4 b)

Other Ways of Switching on Transistors
Page 54 1 b) 2 c) 3 c)

Transducers and Relays
Page 57 1 b) 2 f) 3 g) 4 h) 5 c)
6 d) 7 a) 8 i) 9 e)

Fault Finding
Page 59 Because they can be up to 10% different from the marked values
Page 60 1 a) 9 V b) 9 V c) about 0.7 V – the LED is alight so the transistor is conducting
2 The fault will be a poor connection between A and D or between B and E, but it could also be a broken connection between both
3 a) good b) faulty c) good d) good e) faulty
  f) faulty g) faulty h) good i) good j) good
  k) faulty

Amplifiers
Page 63 1 a) 2 a) 3 d) 4 b)

Integrated Circuits
Page 65 1 c) 2 b)

Logic Systems
Page 69
  A B C D E
  0 0 0 0 1
  1 0 0 0 1
  0 1 0 1 0
  1 1 1 1 0
Page 70 1 a) 2 b)

Microcomputer Systems
Page 73 1 a) 2 b) 3 b) 4 d)

Audio Systems
Page 76 1 d) 2 b) 3 e) 4 c)

Radio Receiver Systems
Page 79 0.5 Hz, 0.25 m
Page 80 1 f) 2 d) 3 c) 4 d)

Feedback Control Systems
Page 83 1 d) 2 b) 3 d)

Sources of Danger
Page 87 1 d) 2 a) 3 b) 4 e)

Fuses and Residual Current Devices
Page 88 a) 5 A b) 2 A c) 13 A d) 13 A
Page 90 1 c) 2 d) 3 b) 4 a)

Power Supplies, Cable Sizing, Device Handling
Page 91 U e) V c) W b) X f) Y a) Z d)
Page 93 1 d) 2 d) 3 c) 4 a)